SURFACE MOUNTED ASSEMBLIES

SURFACE
MOUNTED ASSEMBLIES

Edited by
J. F. PAWLING

ELECTROCHEMICAL PUBLICATIONS LIMITED
1987

ELECTROCHEMICAL PUBLICATIONS LIMITED
8 Barns Street, Ayr, Scotland

ISBN 0 901150 21 5

Typesetting by Brian Robinson, Buckingham
Printed in Great Britain by The Anchor Press Ltd
and bound by Wm Brendon & Son Ltd,
both of Tiptree, Essex

PREFACE

Mullard Ltd is the electronic component arm of Philips in the United Kingdom. Mullard Mitcham has existed for almost sixty years as a development and manufacturing site for vacuum and gas-filled devices, but also, during that time, has been involved in semiconductor manufacture (1955–67), Thin and Thick film hybrids since 1958, computer core-based memories between 1958 and 1975, and electronic sub-assemblies based on printed circuit boards for a range of applications, varying from Consumer to Military, since 1967. It was against this background that the Electronic Assemblies Division at Mullard Mitcham decided in 1980 to invest in Surface Mounted Assembly Technology. This resulted, with the assistance of the Central Materials Laboratory situated on the same site, in the setting up of a team of engineers and technologists to introduce the new processes, to design the products, and commission the production machinery. This book is written by members of that team, experts in a variety of different fields, who combine theory with many years of practical industrial experience.

Surface Mounted Assembly Technology has been born out of a marriage of Hybrid Thin and Thick Film techniques; and Printed Circuit Board Assemblies. The result is a very cost-effective means for miniaturisation, plus greater design freedom than before. These attractive properties have resulted in a very fast exploitation of Surface-Mounted Assemblies over a very broad spectrum of applications from military to consumer. It is a subject that needs to be well understood by those working in Electronics in both R&D and production, such as system and circuit design engineers, technologists, process and production engineers.

This book has been written with the objective of providing a full, and yet concise, coverage of all the major aspects of the application of Surface Mounted technology. It is hoped that, while emphasising the practical aspects for the benefit of those already working in industry, it will be a useful introduction for those studying to become professionally qualified. References, and sometimes a bibliography, are provided at the end of each chapter to enable the reader to extend further and deepen his understanding on specific aspects of the subject.

I must thank all those who have contributed as authors of this book and, in particular, Ges Willard, who has further assisted in many discussions with myself in the reading of drafts and in generating many of the illustrations. We are particularly grateful to our many colleagues within Philips who have, sometimes unknowingly, provided much help and assistance. In particular, I would like to mention Ron Wildy for his assistance with Chapter 7 and Mrs Jean Baker, without whose patient typing and retyping this book would never have been completed.

Finally, I must thank the Mullard company for its encouragement and support, but must emphasise that all statements and opinions expressed in this book are those of the individuals themselves and not those of the Mullard Company.

John Pawling
November, 1986

BIOGRAPHIES

The Editor

After gaining his BSc in Physics and Mathematics at Leicester University, and completing National Service in the Royal Signals, **John F. Pawling** undertook a two-year graduate training scheme with Mullard. In 1957 he joined the Mullard Application Laboratories where he contributed to the early work on the use of semiconductors in audio, radio, television, DC converters and oscilloscopes. In 1967 he left to help in the management of a new activity to design and manufacture electronic sub-assemblies for a wide spread of application areas and using a variety of technologies including both Thin and Thick Films, and Printed Circuit Boards. He has been instrumental from 1980 onwards in the setting up of a Surface Mounted Assemblies activity at Mullard Mitcham. Mr Pawling is a Fellow of the Institution of Electronic and Radio Engineers.

The Contributors

S. G. Mullett has worked in the fields of plastics and polymer technology for twenty-four years. While with Yarsley Research Laboratories he gained experience with polymer development and processing, printing inks, adhesives, and epoxide resin technology. He studied at the Polytechnic of the South Bank, London, to become a graduate of the Plastics and Rubber Institute. In 1970 he joined the Central Materials Laboratory at Mullard Mitcham as a plastics technologist. He is currently involved with many aspects of polymer technology and, in particular, with adhesives, surface coatings and encapsulants for the electronics industry.

After graduating from King's College, London, **G. H. Snashall** undertook six years commissioned service in the Royal Air Force. He joined Mullard in 1954 as production engineer working on UHF transmitting tubes. He was appointed Head of Transmitting Tube Development in 1960 and of the combined Gas Discharge and Transmitting Tube Development Department eight years later. He transferred in 1972 to the Electronic Assemblies Division as Technology Manager. Three years later he led a team developing microchannel plates and fibre components for passive night viewing devices. He transferred in 1980 to the Central Materials Laboratory as Group Leader for advanced device development including substrates using Thick and Thin film techniques.

After completing an initial training programme involving an introduction to all aspects of the design and production of electronic components, **G. A. Willard** joined the Applications Laboratory at Mullard Mitcham and worked primarily on semiconductor applications in the fields of Audio, Radio and general industrial control. He transferred in 1967 as a development engineer to a new activity responsible for the development of electronic assemblies for the Entertainment and Domestic markets. He later became involved with the Automobile and Professional markets and in 1980 was appointed to lead the team responsible for introducing Surface Mounted Assemblies technology into the Mullard activity at Mitcham. He is a member of the Institute of Electronic and Radio Engineers.

Edmund George Evans, BSc (London), FISTC, has lectured in electronics, taught physics, and run the Mullard technical publications unit. He was Editor of 'Mullard Technical Communications' and first Editor of the international Philips journal 'Electronic Components and Applications'. He now offers a comprehensive technical authorship service at Evans Documentation Services.

Keith R. Stone graduated from Leeds University in 1980 with a degree in Materials Science. His first appointment was as a Research Officer at the International Tin Research Institute in the Soldering Department. In 1982 he took up the position of Development Engineer and later Chief Technologist at Mullard Mitcham in the Electronic Assemblies Division, where he worked on the development of new electronic assembly processes and technologies. Mr Stone has since moved to Technophone Ltd, Woking, as Production Engineer.

CONTENTS

PREFACE   v
BIOGRAPHIES   vii

CHAPTER ONE
The Beginnings of Surface Mounted Assemblies   1
J. F. Pawling

1.1   General   1

1.2   The Thermionic Valve Era   1

1.3   The Printed Circuit Board   3

1.4   The Inserted Component   4

1.5   Passive Components   4
1.5.1   Resistors   5
1.5.2   Capacitors   5
1.5.2.1   Ceramic Capacitors   5
1.5.2.2   Electrolytic Capacitors   6
1.5.2.3   Foil Capacitors   7

1.6   Active Components—The Transistor   8
1.6.1   The Silicon Integrated Circuit   8

1.7   Hybrids   9
1.7.1   Thin Film Circuits   9
1.7.2   Thick Film Circuits   9

1.8   The First Surface-mountable Devices   10

1.9   Inductors   11

1.10   Variable Resistors and Capacitors   12

1.11   Fixed Resistors   12

1.12   The Arrival of SMA   13

CHAPTER TWO
Substrates   15
G. H. Snashall and J. F. Pawling

2.1   Introduction   15

2.2   General Requirements for Surface Mounting Substrates   15
2.2.1   Compatibility of Substrate and Components   16
2.2.2   Compatibility with the Machine   17
2.2.2.1   Locating Methods   17
2.2.2.1.1   Location on the Edge   17
2.2.2.1.2   Mechanical Location   18

2.2.2.1.3	Location Accuracy	18
2.2.2.2	Uniformity of Thickness	18
2.2.2.3	Flatness or Bow	18
2.3	Types of Substrates	19
2.3.1	Ceramic Substrates	19
2.3.1.1	Alumina Substrates	19
2.3.1.2	Beryllia Substrates	21
2.3.1.3	Aluminium Nitride	22
2.3.1.4	Silicon Carbide/Beryllia Composite	22
2.3.1.5	Size of Ceramic Substrates	23
2.3.1.6	Holes	23
2.3.1.7	Metallisation Methods for Ceramic Substrates	24
2.3.1.7.1	Thin Film Technology	24
2.3.1.7.2	Thick Film Technology	25
2.3.2	Organic Substrates	27
2.3.2.1	Properties of PCB Materials	28
2.3.2.2	Additive Processes	28
2.3.2.3	Conductor Dimensions	29
2.3.2.4	Outline Shaping and Hole Generation	29
2.3.2.5	Multilayer PCBs	30
2.3.2.6	Chipstrates	30
2.3.2.7	Flexible Circuits	31
2.3.2.8	Additional Processes	31
2.3.2.9	Polymer Conductor Resistor Materials	32
2.3.3	Metal-cored Substrates	32

CHAPTER THREE
Passive Surface Mounted Devices   34
G. A. Willard and E. G. Evans

3.1	Introduction	34
3.2	Surface Mounted Devices—General Characteristics	36
3.3	Packaging Surface Mounted Devices	37
3.4	Resistors	39
3.4.1	Manufacture of SM Resistors	40
3.4.2	Quality, Reliability and Performance of SM Resistors	42
3.4.2.1	Zero-hour Quality at Delivery	42
3.4.2.2	Reliability in Operation	42
3.4.2.3	Solderability	43
3.4.2.4	Temperature Coefficient	43
3.4.2.5	Voltage Coefficient	44
3.4.2.6	Other Characteristics of SM Resistors	44
3.4.3	Cylindrical Leadless MELF Resistors	45
3.4.4	SM Resistors with Higher Power Dissipation Ratings	46
3.4.5	Thick Film SM Resistor Networks	46
3.5	SM Ceramic Multilayer Capacitors	47
3.5.1	Manufacture of SM Ceramic Multilayer Capacitors	48
3.5.2	Dielectrics in Ceramic Capacitors	50
3.5.2.1	Temperature Coefficients	52
3.5.2.2	Tolerances on Temperature Coefficients	52
3.5.2.3	Operating Temperature Ranges	52
3.5.2.4	Dielectric Strength	53
3.5.2.5	Insulation Resistance	53
3.5.2.6	Dielectric Losses	53
3.5.2.7	Ageing	53
3.5.2.8	Voltage Ratings	54
3.5.3	Quality and Reliability of CMCs	54

3.5.4   High Value and High Voltage Ceramic Capacitors   54
3.5.5   Mica Chip Capacitors   54

3.6   Cylindrical MELF Ceramic Capacitors   54

3.7   Metallised Polyester SM Capacitors   55

3.8   Aluminium Electrolytic SM Capacitors   56

3.9   SM Solid Tantalum Capacitors   57

3.10   Miniature SM Inductors   59

3.11   Adjustable SM Inductors and Transformers   62

3.12   Ceramic Filters   62

3.13   Crystals and Oscillators   63

3.14   SM Potentiometers   65

3.15   Trimmer Capacitors   66

3.16   Thermistors   67
3.16.1   NTCs   67
3.16.2   PTCs   67

3.17   Connectors   68
3.17.1   Flex-rigid Prints   71
3.17.2   Elastomeric Connectors   72
3.17.3   Sockets   73
3.17.4   Switches   73
3.17.5   Keyswitches and Keyboards   75

3.18   Availability of SM Passive Components   75

CHAPTER FOUR
Active Surface Mounted Devices   78
E. G. Evans and G. A. Willard

4.1   Introduction   78
4.1.1   Unencapsulated Chips   79
4.1.2   Mixing Technologies in Mixed Prints   81

4.2   Active Surface Mounted Devices—General Features   81

4.3   Discrete SM Semiconductors   82
4.3.1   The SOT-23 Encapsulation   83
4.3.2   The SOT-89 Encapsulation   84
4.3.3   The SOT-143 Encapsulation   85
4.3.4   Cylindrical Diodes   87
4.3.4.1   The SOD-80 Diode Encapsulation   87
4.3.4.2   The MELF Diode Encapsulation   89
4.3.5   Conventional Outlines Adapted for SMDs   89
4.3.6   Performance Obtainable with Discrete Semiconductor SMDs   90
4.3.7   Soldering Semiconductor SMDs   90
4.3.8   Packaging Semiconductor SMDs   91
4.3.9   Reliability of Discrete Semiconductor SMDs   92
4.3.9.1   Zero-hour Quality and FITs   93

4.4   Surface Mounted Integrated Circuits   94
4.4.1   Encapsulations for Surface Mounted ICs   95

4.4.1.1	Small Outline Integrated Circuits	95
4.4.1.2	Chip Carriers	98
4.4.1.3	The Quad Flatpack	100
4.4.1.4	Pad Array and Leadless Grid Array Encapsulations	102
4.4.1.5	Other Surface Mounted Encapsulations for ICs	104
4.4.2	Performance of Surface Mounted ICs	105
4.4.3	Soldering Surface Mounted ICs	105
4.4.3.1	Thermal Cycling and Mechanical Strength	106
4.4.3.2	Sockets for Integrated Circuits	106
4.4.4	Packaging of Surface Mounted ICs	108
4.4.5	Quality and Reliability of SM ICs	108
4.5	Surface Mounted Semiconductor LEDs	109
4.6	Availability of Semiconductor SMDs	111

CHAPTER FIVE
Design of the Circuit Layout

113

J. F. Pawling, G. A. Willard and K. R. Stone

5.1	Introduction	113
5.2	Design Options	113
5.3	Design Restrictions	116
5.3.1	CAD System	116
5.3.2	Production Machinery	116
5.3.2.1	Constraints Imposed by Placement Machinery	116
5.3.2.2	The Influence of the Soldering Method	118
5.3.2.3	Influence of Testing Requirements	121
5.4	General Design Rules	121
5.4.1	Guidelines	121
5.5	Computer-aided Design of the Substrate Layout	125
5.5.1	General Requirements	125
5.5.2	PCBs with Inserted Components	126
5.5.2.1	CAD Equipment	126
5.5.2.2	CAD Software	127
5.5.2.3	Steps in a Layout Design	127
5.5.3	SMA and Mixed Technology Layouts	128

CHAPTER SIX
Component Placement

138

K. R. Stone

6.1	Introduction	138
6.2	Hand Assembly	138
6.3	Machine Assembly	139
6.3.1	Machine Functions	140
6.3.1.1	Substrate Positioning and Feed	140
6.3.1.2	Adhesive or Solder Paste Application	140
6.3.1.3	Parts Feed	141
6.3.1.4	Pick-up Heads	142
6.3.1.5	Programming	142
6.3.1.6	Component Safety Features	143
6.3.1.7	Curing System	143
6.3.2	Machine Types	144
6.3.2.1	Semi-automatic Assembly Aids	144

6.3.2.2	Sequential Pick and Place Machines	144
6.3.2.3	Software Programmed Multiple Head Machines	144
6.3.2.4	Hardware Controlled Machines	144
6.3.3	Survey of Some Popular Machines	145
6.3.3.1	Semi-automatic Hand Assembly/Training Aids	145
6.3.3.2	Sequential Pick and Place Machines	147
6.3.3.3	Sequential Pick and Place Machines (Medium-high Volume)	152
6.3.3.4	Software Programmed Multiple Head Machines	157
6.3.3.5	Hardware Controlled Machines	159

CHAPTER SEVEN
Component Attachment 163
J. F. Pawling

7.1	Introduction	163
7.2	Overview of Attachment Methods	163
7.3	Reflow Methods	163
7.3.1	Selection of the Solder Cream	165
7.3.2	Application of the Solder Cream	165
7.3.2.1	Pin Transfer	165
7.3.2.2	Pressure Dispensing	166
7.3.2.3	Screen Printing	166
7.3.3	Component Placement and Predrying	167
7.3.4	Reflow Heating Methods	167
7.3.4.1	Conduction Methods from Below	167
7.3.4.2	Heating from Above	168
7.3.4.2.1	Infra-red Soldering	168
7.3.4.2.2	Laser Soldering	169
7.3.4.2.3	Hot Gas Soldering	169
7.3.4.3	Vapour Phase Soldering	169
7.3.4.4	Resistance Heating	171
7.3.4.5	Summary of Reflow Soldering	172
7.4	Glue and Solder Attachment	173
7.4.1	Selection of the Adhesive	173
7.4.2	Glue Dispensing	174
7.4.3	Handling after Adhesive Dispensing	175
7.4.4	Machine Flow Solder Methods	176
7.4.4.1	Dip Soldering	176
7.4.4.2	Drag Soldering	177
7.4.4.3	Wave Soldering	177
7.4.4.4	Single Jet Wave	177
7.4.4.5	Double Wave Systems	178
7.4.4.6	Hot Air Knives	179
7.4.4.7	Orientation of Components	180
7.4.4.8	Summary of Flow Solder Methods	181
7.5	Hand Soldering	181
7.6	Conductive Glues	182
7.7	Post-Cleaning	182

CHAPTER EIGHT
Post Attachment Processes 185
J. F. Pawling and S. G. Mullett

8.1	Introduction	185
8.2	Inspection	185

8.2.1	Substrate Inspection	186
8.2.2	Solder Paste Deposition	186
8.2.3	Pre-solder Inspection	187
8.2.4	Solder Inspection	189
8.3	Electrical Testing	192
8.3.1	In-circuit Component Testing	192
8.4	Fault Finding and Repair	194
8.5	Protective Coatings and Encapsulations	195
8.5.1	Conformal Coating	196
8.5.1.1	Dipping Techniques	196
8.5.1.2	Spraying Techniques	196
8.5.1.3	Other Techniques	197
8.5.1.4	Materials for Conformal Coating	197
8.5.2	Solids-based Conformal Coating	197
8.5.3	Encapsulation	198

CHAPTER NINE
The Practical Advantages of SMA   199
J. F. Pawling and G. A. Willard

9.1	Overview	199
9.2	Miniaturisation	199
9.3	Greater Design Freedom	203
9.4	Reduction in Complexity	206
9.5	Comparison with Thin and Thick Film Hybrids	206
9.6	Unusual Shapes	208
9.7	'Second Thoughts' Designs	209
9.8	Tabular Comparison	210
9.9	Cost Comparisons	210
9.9.1	Materials	212
9.9.2	Direct Labour	212
9.9.3	Capital Depreciation	213
9.9.4	Design Costs	213
9.10	Reliability	214
9.10.1	Testing for Reliability	214
9.10.2	What Problems to Expect	215
9.10.3	Thermo-mechanical Reliability Test Results	216
9.10.3.1	Thermal Cycling Results	218
9.10.3.2	Flexure Cycling Results	220
9.10.3.3	Electrical Results	220
9.10.3.4	Interpretation of the Results Obtained	220
9.10.3.5	Conclusions Drawn from the Practical Tests	225
9.10.4	Summary	225

CHAPTER TEN
A Practical Approach 227
J. F. Pawling and G. A. Willard

10.1 The Product Life Cycle 227

10.2 Market Requirement and Detailed Specification 228

10.3 Reasons for Selecting SMA 229

10.4 The Design Stage 229
10.4.1 Component Selection 230
10.4.2 Mechanical Constraints 231
10.4.3 Board Layout Stage 231
10.4.3.1 Layout Design 233

10.5 Design Evaluation 233
10.5.1 Stages of Sample Manufacture 234
10.5.1.1 Stage 1—First Prototype 234
10.5.1.2 Stage 2—Customer Evaluation Devices 235
10.5.1.3 Stage 3—Life Test Samples 235

10.6 Manufacture and Test Stages 236
10.6.1 Screen Print Solder Cream 236
10.6.2 Top Surface SMD Placement 238
10.6.3 Reflow Soldering 238
10.6.4 Underside SMD Placement 239
10.6.5 Leaded Assembly 240
10.6.6 Flow Soldering 241
10.6.7 Testing 241
10.6.8 Summary 242
10.6.9 Monitoring of the Processes and Reaction to Change 242

10.7 Field Performance 243
10.7.1 Reliability 243
10.7.2 Repair 244

10.8 Product Upgrading 244

CHAPTER ELEVEN
The Future 246
J. F. Pawling and G. A. Willard

11.1 The Future 246

11.2 Greater Packing Densities 246
11.2.1 Two Dimensions 246
11.2.1.1 The Substrate 246
11.2.1.2 The Components 249
11.2.2 Three Dimensions 249
11.2.3 Heat Dissipation 253
11.2.3.1 Metal Substrates 253
11.2.3.2 Separated Electrical and Thermal Conduction 253

11.3 Design Considerations 255

11.4 Summary 255

AUTHOR INDEX 256
SUBJECT INDEX 257

Chapter 1

THE BEGINNINGS
OF SURFACE-MOUNTED ASSEMBLIES

J. F. PAWLING
Mullard Mitcham, Mitcham, Surrey, England

1.1 GENERAL

The evolution of Surface-mounted Assemblies results from three quarters of a century of electronics history. It could have been earlier. Sometimes the partial failures of alternative manufacturing technologies caused setbacks, but the enormous success of the printed circuit board and its inserted components diverted effort that could have been employed on developing surface mountable components and consequential assembly technologies.

There is a close relationship between the design of electronic components and the processes used to assemble them into a working circuit. Many proposals for the assembly of electronics have foundered on the non-availability of the right components at an economic price. In turn, it is also true that the direction taken by component developments is determined by the current capability of the technologies used in their manufacture. There must additionally be one other ingredient for progress and that is a market pull or a firm need for change to take place. How much slower might have been the development of electronics without the American space programme demanding miniaturisation and immense complexity within the same application.

1.2 THE THERMIONIC VALVE ERA

The strongest component influence in electronics has been the active device, the one capable of amplification and hence able to change the condition of an input signal. Carbon composition resistors and aluminium electrolytic capacitors were first made in the 1880s, but electronics could only take off with the birth of the thermionic triode valve in 1906. For the next fifty years the valve reigned supreme and, as befits a king, the other components populating electronic systems moulded themselves to its wishes.

It is interesting to note that the concept of integration has been with us for a long time, and as early as 1926[1] electronic engineers were attempting to offer a complete circuit within one device. Figure 1.1 shows three triodes coupled by resistors and capacitors all within the one glass envelope, a device announced at the German Wireless Exhibition in Berlin with a statement by the makers that they were to discontinue the manufacture of ordinary single valves. They had the vision but not the right technology.

The thermionic valve was relatively expensive, had a limited working life, and, although it became smaller during its latter years, was always bulky. It was considered to be a pluggable component that could be replaced when necessary. As the number of electrodes increased, so did the connecting pin count and a typical valve base socket had eight or nine sockets in a ring of 12·5 mm diameter or more. Sufficient inter-pin spacing was necessary to ensure adequate insulation remembering that valves generally operated at what are often now considered to

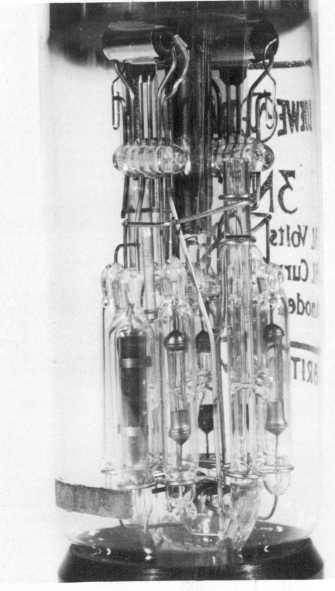

Fig. 1.1   An early example of integration. Three triode valves coupled by resistors and capacitors all within one glass envelope.

be high voltages. The mechanical and physical requirements to ensure a strong and reliable piece of equipment demanded that the valve sockets were mounted on an aluminium or cadmium plated steel chassis and bolted down. Circuit designers, radio hobbyists, early hi-fi audio enthusiasts, and many others, spent hours of preparation cutting, bending, drilling, and in other basic chassis work before the assembly of circuit components could begin. Once the chassis was completed, the large components such as transformers and electrolytic capacitors were bolted into place and finally the remaining components were assembled. Resistors, capacitors and other components were now required to use one of two methods of construction. If sufficiently large, they would also have brass screw connection terminals and feet that could be bolted to the chassis but, if light and small enough, they provided the dual rôle of passive component and interconnection by having wire terminations built into their bodily construction. These wires were wrapped around tags on the valve bases or on tag strips and then soldered using a hand held soldering iron. Figure 1.2 shows a typical valve chassis assembly.

Fig. 1.2 A typical valve chassis assembly – for a radio receiver. Wire-ended components are strung in three dimensions and are soldered to valve bases and tag strips.

1.3 THE PRINTED CIRCUIT BOARD

Two things occurred soon after the second world war, which together broke the dominance of the valve and metal chassis. They were the announcement of the transistor in 1948 and the advent of the printed circuit board. The combination of the much smaller and more efficient transistor together with the two dimensional concept of the circuit board accelerated the wider application of electronics, but at the same time emphasised the need for wire-ended components and may therefore have delayed the birth of Surface-mounted Assemblies. Many workers in the 1940s were advocating an integrated manufacturing approach which later generations recognise as Thick and Thin Film hybrid technologies. An example was J. A. Sargrove who, in a paper given in 1947[2], refers to his aim 'to give electronic apparatus a form which lends itself to fully mechanised production methods, a principle has been evolved of treating a circuit, not as an assembly of component elements but as a compound whole.' He went on to say, 'It will be appreciated, of course, that components such as valves, loudspeakers and electrolytic capacitors cannot be applied in this manner.' The manner he was referring to was the deposition of conductors, resistors, inductors and capacitors. The ideas were there but again limited by availability of components and technologies.

1.4   THE INSERTED COMPONENT

Printed circuits rapidly became the construction method for all electronics unless constrained by high frequencies or high powers. The printed circuit substrate materials such as phenolic, epoxy or polyester impregnated fibreglass or paper were relatively soft and holes could be punched or drilled in them. The easiest assembly method was to use a wire-ended component, insert the wire through a hole, bend it over to provide temporary retention and solder the wire to a ring of copper around the hole on the opposite side of the board to the body of the component (see Figure 1.3). Because this was the easiest assembly method,

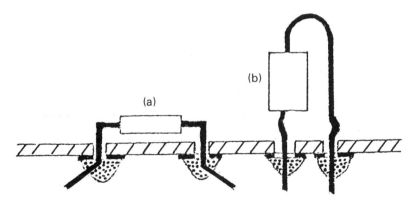

Fig. 1.3   Axial and Radial component leads inserted through holes in the printed circuit board and soldered to copper on the reverse side. (a) Axial component with leads clinched outwards. (b) Radial component with preformed leads to space the body off the board surface.

whether using hand insertion or automatic insertion machines, there was a market pull for all components to meet this requirement. This meant, in many cases, smaller and less heavy components than had previously been acceptable with a metal chassis. Fortunately, new materials and processes were becoming available to enable component designers to meet the new challenge.

1.5   PASSIVE COMPONENTS

Prior to the second world war the main application for electronics was in domestic radio receivers. The radio receiving valve had the fortunate property that it could accept wide initial value tolerances and poor stability of its surrounding passive components. An expansion in applications demanded better components.

1.5.1 Resistors

Early resistors were of the carbon composition type where a granular mixture of carbon or graphite, resin and refractory filling was pressed into a tubular container, cured and then fitted with metal caps on both ends. As there was no method of adjusting the resistance value afterwards, a tolerance of 20%, about the nominal value, was normal. An alternative manufacturing method was to wind high resistance wire on a tubular ceramic former to produce the so-called wire wound resistor used for lower resistance values and for higher power dissipation—which fortunately tended to go together in many applications.

Film resistors followed commercially soon after the second world war. In all cases a film is deposited on a ceramic rod and a spiral track is cut through the resistive layer to bring it up to the required value of resistance (see Figure 1.4).

Fig. 1.4 A typical film resistor in which a spiral track is cut through the resistive layer to bring it up to the required value.

Cracked carbon film resistors were first on the market and in fact had been available in Europe since the early 1930s, and in the 1950s metal oxide and metal film resistors also entered the field. Resistors were now much more sophisticated. Close tolerance initial values, high stability and small changes with temperature could be offered. Concurrently with these advances in the discrete resistors came the development of integrated resistor technologies such as thick and thin film, which will be dealt with later in this chapter.

1.5.2 Capacitors

For many years circuit designers could choose between three types of capacitor—wet aluminium electrolytics for the highest values of capacitance; wound paper and metal foil with their inherently high self-inductance; and for low inductance, high Q applications such as tuned circuits there were capacitors made from sheets of mica either interleaved with metal foil or silver metallised.

1.5.2.1 CERAMIC CAPACITORS

Ceramic as a dielectric became an additional choice first in Germany[3] in 1933. It was cheaper than mica and had the common advantage of low self-inductance compared with film and foil. By 1953 large numbers of tubular ceramic capacitors were being used, but although it was realised that, stacked flat, thin plates of ceramic would provide a significant size reduction, it was not considered economically feasible at that time.

Because the capacitance of a parallel plate capacitor is proportional to the relative permittivity of the dielectric used and inversely proportional to its thickness, it follows that larger capacitances in a given volume can be obtained by either reducing the thickness or using dielectrics with larger relative permittivities[4]. During the first 15–20 years ceramic capacitors employed materials with permittivities up to 100. The discovery of the ferro-electric properties of barium titanate in 1943 brought materials with a permittivity of 1000 and more within reach.

In the 1960s the problems of reducing the thickness below 200 microns were solved. Flat plate ceramic capacitors are produced by pressing, extruding and calendering, but below 200 microns thickness the plates become fragile and difficult to handle. As modern dielectrics can withstand electric field strengths of 2 to 3 volts per micron, so a 50 volt rated capacitor requires a thickness of only 20–25 microns. Such a device would clearly have 8–10 times the capacitance of a 200 micron thick plate. Two ways of combining the required mechanical robustness and electrical thickness became possible. One was a barrier layer in which a dielectric thickness of a few microns is achieved with a physical thickness of a few hundred microns and the second type was a ceramic multilayer consisting of a body with embedded electrode layers which are alternately connected to the metallised end terminations so that individual capacitors formed by the layers are connected in parallel (see Figure 1.5).

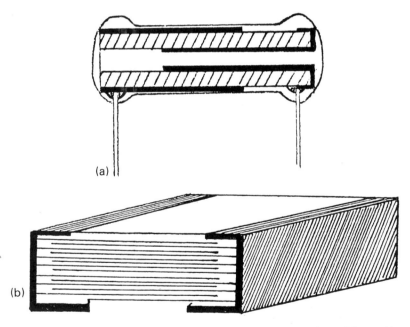

Fig. 1.5   Two types of ceramic capacitor. (a) Tubular, and (b) Multilayer chip.

1.5.2.2   ELECTROLYTIC CAPACITORS

The wet aluminium electrolytic capacitor has existed for a century or more. Whilst offering the largest capacitance to volume ratio, these were bulky, and also suffered from premature failures due to the aggressive action of the electrolyte on the oxide layer forming the dielectric, particularly when not used for long periods. To extend the life expectancy circuits were designed to operate the capacitors well below their maximum rated voltage. This only meant an even bulkier capacitor being used for a given application. The advent of the etched aluminium foil and new low water content electrolytes have done much to improve this type of capacitor[5]. Etching factors of approximately 100 for modern low voltage capacitors enable the volumetric efficiency to be increased by a proportional amount.

The wet aluminium electrolytic was followed by the tantalum, and more recently by the solid aluminium electrolytic. Both offer even smaller sizes but acceptance of tantalum for more cost-conscious applications has been limited by the high and fluctuating price of the raw material. Solid tantalum capacitors have been available since the early 1960s and have been used because of their small size in thin and thick film hybrid applications. Typically a 10 microfarad solid tantalum capacitor will take up less than 10 sq mm of substrate area, while a wet aluminium electrolytic of the same electrical value may require three times this.

1.5.2.3 FOIL CAPACITORS

The second world war, as was so often the case, provided the stimulus for rapid changes in foil capacitors. The construction of a wound film and aluminium foil capacitor is illustrated in Figure 1.6. Humidity effects underlined the weakness of

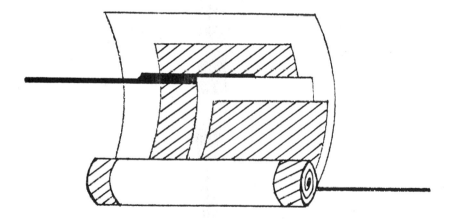

Fig. 1.6 The construction of a film and foil wound capacitor. Two or more insulating films are used as the dielectric and another film layer is used as an outer insulation. External wire leads are inserted to rest alongside and make electrical contact with the aluminium foil electrodes.

the earlier waxed tubular paper dielectric capacitor. Metal cased tubular types with neoprene seals proved more reliable but not smaller or cheaper. Metallisation of the dielectric in place of metal foil offered some space saving, sometimes of the order of 2–4 times. Modern plastics, such as polycarbonate, polystyrene, polypropylene and polyester, began to be introduced in the 1950s, and replaced paper for many applications. Being thin, yet strong, they enabled smaller capacitors to be made. Today, wound film capacitors are small compared with 40 years ago, but are nevertheless still relatively bulky for modern miniaturised electronics. A further problem is that many of the plastic films will deform if subjected to temperature close to that of molten solder, a requirement for many modern assembly processes. In spite of this some manufacturers now offer surface mountable film capacitors.

1.6   ACTIVE COMPONENTS—THE TRANSISTOR

Returning now to the active components in electronics, the development of the transistor during the 1950s coincided, as already stated, with the initial use of the printed circuit board in production. Earliest devices followed valve making techniques with glass encapsulations and wire connections from the base of the component. Later products were in metal cases followed in due course by transfer-moulded plastic encapsulations, but all with wire leads of a rather flexible material and hence indefinable positions. The intention was a product for printed circuit board insertion, but in practice this meant hand insertion. During the 1970s Mullard attempted to solve the assembly problem by introducing a tagged version, the so-called Lockfit transistor (see Figure 1.7), but it never became sufficiently popular.

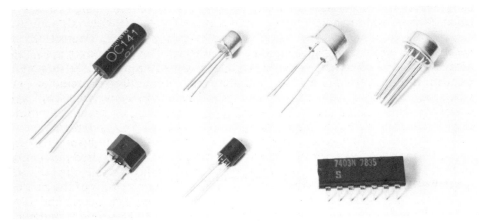

Fig. 1.7 A selection of semiconductor encapsulations. For discrete devices, black painted glass, metal T018 and T05, plastic T092 and Lockfit. For integrated circuits, multi-pin T099 and plastic Dual-in-Line.

1.6.1   The Silicon Integrated Circuit

The introduction of silicon planar technology in the 1960s enabled the integrated circuit to become a reality. Although the earliest devices were available in wire-ended metal cans, the Dual-in-Line package in either ceramic or plastic, first introduced in the mid 1960s (Figure 1.7), became a popular international standard, and by 1980 accounted for 85% of encapsulated integrated circuits. Connection to the world outside the integrated circuit was made by two well defined rows of tags at 0·1 inch pitch, thus reducing the problems of getting sixteen or so flexible wires through their respective holes at the same time! The packaging of both the transistor and integrated circuit were following the dictates of the printed circuit board. Furthermore, in order to ensure the greatest usage, the semiconductor packages were chosen so that all equipment makers could use them, i.e., the lowest common denominator of technological ability. This approach prevented closer pin spacings and smaller packages being introduced, yet there was an increasing demand for miniaturisation of electronic equipment and greater packing densities. This pressure, mainly for military and space applications, encouraged the introduction of Thin and Thick Film hybrid technologies.

1.7 HYBRIDS

1.7.1 Thin Film Circuits

Sargroves' 1947 paper[2] visualised the printing of resistors, inductors and capacitors. Brunetti and Curtis[6], in the following year in the Proceedings of the Institution of Radio Engineers, presented a very detailed outline of the alternative ways of producing complete circuits. Initially the three technologies of Thin Film, Thick Film and Silicon Integration were seen as competing and their advocates claimed that each was capable of doing the total circuit function. It was only later that silicon integration was recognised as being complementary to the two film technologies.

Thin film microcircuits began to appear at the same time as the transistor. They are prepared by boiling material under reduced pressure or using a high voltage to cause gas ionisation and hence ion bombardment of a source material, and then allowing the resulting vapour to be deposited onto the substrate. Deposition can be limited to a selected area using a mask or alternatively can be over the whole area followed by selectively etched removal from unwanted parts of the substrate. Conductors are usually of gold and resistors of nichrome. It was predicted that conductors, resistors, capacitors, inductors and eventually active transistor type devices would all be made by the one technology. In time, the advantages of thin films were seen to be very narrow conductor line widths, and high quality resistors capable of being accurately adjusted to value. The disadvantages were that thin film capacitors were limited to a maximum of a few hundred picofarads and of low Q and, as only one film resistivity was generally available, the range of resistor values was necessarily limited by the allowable aspect ratios of the printed resistor patterns. Most capacitors, inductors and all the semiconductors had to be mounted on the glass or ceramic substrate and, unlike the relatively soft materials of a printed circuit board, holes were not easily introduced for insertion of the wire leads of the readily available cheap components. The alternative possibilities were to utilise the components in a way for which they were not designed by forming feet with the wire terminations, then standing them on a conductor pad and reflow soldering them, or to seek new specialised components. This was a problem shared with thick film circuits which, although first used by Centralab in 1941 for a proximity fuse, were commercially introduced over a wider application range in the 1960s.

1.7.2 Thick Film Circuits

Thick Film technology overcomes the limited resistor value range of Thin Film by using an extensive range of pastes of differing resistivities which are screen printed onto the substrate, then dried and fired. The conductors, because they are printed in a similar manner to the resistors, cannot quite match the Thin Film techniques in producing fine line geometries. Examples of Thin and Thick Film resistor patterns are shown in Figure 1.8. Thick Film technology is generally considered to be the cheaper for production, both in initial investment and material costs. However, it is essentially a method of producing resistors and conductors and most other components have to be added afterwards. In the early formative years of the 1950s and '60s, the choice was between cheap wire-ended devices or relatively expensive special components. They were expensive because

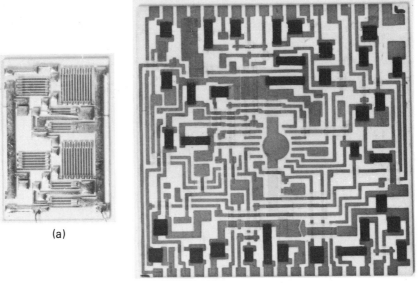

Fig. 1.8 Examples of Thin and Thick Film Hybrid Substrates. (a) Thin Film on glass. (b) Thick Film on Alumina.

the market was small compared with the quantities used in printed circuit boards. Nevertheless, components for surface mounting began to be developed.

1.8   THE FIRST SURFACE-MOUNTABLE DEVICES

By the later 1960s the film technologies described above were enabling electronic circuits to be executed in a size an order smaller than those using conventional components mounted on a printed circuit board. The deposited resistor occupied less substrate area than its conventional counterpart and was definitely less tall! Applications were generally limited to those where customers were prepared to pay a premium for miniaturisation. Tantalum electrolytics were employed because of their high CV to volume ratio and often semiconductors were used in their bare silicon chip form. Hybrid manufacturers bought partly tested silicon die, back bonded the chips directly onto their alumina or beryllia substrate, and then wire-bonded from the silicon chip to conductor pads on the substrate. Size reduction was achieved but the process was usually considered too expensive for really large scale production quantities and there was often the debate as to whose problem it was when the chip apparently failed to work properly after bonding – the chip supplier's or the user's! A whole range of proposals to solve the semiconductor packaging problems for hybrids were made including flip chips, beam leads, leadless inverted devices (LID) and tape bonded devices. Two plastic encapsulations for semiconductors began to gain favour, particularly in the more consumer orientated European market – the SOT23 for diodes and transistors and the Small Outline or SO pack for integrated circuits. Both bore a marked resemblance to their 'big brother' encapsulations for PCBs and the SO pack is clearly a scaled-down DIL (see Figure 1.9). The SOT23 was introduced in the late '60s for thick and thin film applications and the SO pack in the early '70s initially for use in electronic watches, and then for all forms of thin and thick film circuits. Flat packs, ceramic chip carriers, and later plastic ones,

Fig. 1.9 Examples of encapsulations for silicon integrated circuits. Dual-in-Line and its scaled down SO version. Flat pack, leadless chip carrier and leaded chip carrier.

both leadless and with leads, augmented the selection of IC package arrangements on offer. Figure 1.9 shows examples of these devices. Thus by the early 1970s the components were arriving to enable Surface Mounted Assemblies (SMA) to take place. Potentially cheap semiconductor packages, plate ceramic and small electrolytic capacitors were a technical possibility. Before closing these thoughts on the historical build-up, it is pertinent, for completeness, to consider inductors; variable resistors and capacitors; and fixed resistors.

1.9 INDUCTORS

In the early days of electronics, air-cored inductors for high frequencies, and iron-cored mains and audio frequency transformers dominated the electronic assembly by their size and weight. Two events have helped to bring about a change. One was technological and the other more of a shift in need. The arrival of ferrite with its wide choice of permeabilities and its ability to be manufactured in a large variety of shapes and sizes enabled much smaller transformers and chokes to be designed. As electronics sought greater miniaturisation and packing densities, so the pot core enabled circuit designers to achieve these requirements without circuit instability from unwanted interactions. An example of a miniature pot core used in an SMA application is shown in Figure 1.10. The second event was a change in circuit design philosophy, resulting in the general swing from analogue to digital circuitry which reduces the requirements for large

Fig. 1.10 A miniature ferrite pot core used in an SMA application and individual trimmer resistors and capacitors.

inductances, and the removal of the need for large transformers from the equipment power supply by the application of switch mode circuit techniques.

1.10 VARIABLE RESISTORS AND CAPACITORS

Variable resistors and capacitors fall into two categories. First, there can be a long term need to be variable as, for example, radio tuning control; and secondly, it may be necessary to have a one-occasion only pre-setting of a value, to be maintained and fixed thereafter. Fortunately the former application requires to be sensibly large for the user to handle and is often mounted separately from the remainder of the circuitry. The latter requirement of a pre-set can be made extremely small and indeed for thick and thin film resistors need only be the same size as any other fixed value resistor being laser cut adjusted to value. The use of plastic film solid dielectrics enables very small discrete trimmer capacitors to be made, and for printed circuit applications, very small stand-alone trimmer resistors are available. These are illustrated in Figure 1.10.

1.11 FIXED RESISTORS

It is sometimes necessary to incorporate resistors deliberately into circuits with large negative or positive temperature coefficients, i.e., NTCs (Thermistors) and PTCs. Even with technologies capable of printing resistors, components requiring such specialised parameters are best added separately. Very small bead thermistors have been available since the 1950s for use with Thick or Thin Film hybrids.

Thick and Thin Film Hybrid technologies are inherently capable of producing their own low temperature coefficient resistors. To expand Surface-mounted Assemblies technology to printed circuit board substrates, an electrically similar component was needed. This was technically easy — just a single resistor printed by thick film techniques on a small ceramic substrate. Once this product was available SMA became possible whatever the chosen substrate material.

1.12 THE ARRIVAL OF SMA

By the late 1970s all of the required surface-mountable components were available. Manufacture of printed circuit boards was well established and, although Thin and Thick Film technologies had shown the way in miniaturisation when used with silicon-integrated circuits, they had been unable to break the price barrier into the full range of applications. A technology was required which offered miniaturisation, great circuit design flexibility and lower manufacturing costs. The use, initially in Japan, of surface-mounted components in combination with wire-ended components for television tuners, car radios and television cameras created a big increase in demand for surface-mounted chip components. Increases in production scale caused their market prices to fall. Both the technical and commercial breakthrough had begun. Figure 1.11[7] shows

IC PACKAGING TRENDS (1982 - 1992)

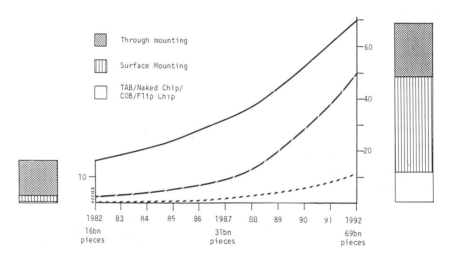

Fig. 1.11 Forecast numbers of through mounting and surface mounting silicon integrated circuits likely to be used in 1992 compared with those used in 1982. (Courtesy of BPA (Technology & Management) Ltd, Dorking, Surrey, England)

the number of Dual-in-Line integrated circuits used with conventional printed circuit board techniques in 1982 compared with the number of surface-mountable silicon devices. It also shows the expected use in 1992. Although the number of DILs has increased, the outstanding change is in the expanding application of Surface-mounted Devices (SMD) such that by the end of the 1980s the number of DILs is overtaken by surface-mountable circuits. Having established what is going to happen, the following chapters will, it is hoped, assist the reader in participating in this revolution in assembly technology.

REFERENCES

1   *Wireless World,* p. 597, November (1926).
2   Sargrove, J. A., 'New Methods of Radio Production', *Journal of the British Institution of Radio Engineers,* **Vol. VII,** No. 1, January-February (1947).
3   Pepper, P., 'Ceramic Dielectrics and Their Application to Capacitors for Use in Electronic Equipment', Proceedings of the Institution of Electrical Engineers, 100 Pt. 11a, No. 3, p. 229 (1953).
4   Hagemann, H. J., Hennings, D. and Wernicke, R., *Philips Tech. Rev.,* **Vol. 41,** No.3, p.89 (1983/84).
5   Otten, Schmickl and Stakhorst, 'Recent Developments in Wet Aluminium Electrolytics', *Electronic Components & Applications,* **Vol. 5,** No. 4, p. 246, September (1983).
6   Brunetti, C. and Curtis, R. W., 'Printed Circuit Techniques', Proceedings of the IRE, **Vol. 36,** p. 121 (1948).
7   Joint Report *Electronic Production*, BPA (UK) Ltd, *Electronic Production*, p. 48, October (1984).

BIBLIOGRAPHY

1   Dummer, G. W. A., 'Modern Electronic Components', Sir Isaac Pitman & Sons.
2   Holmes, P. 'J. and Loasby, R.G., 'Handbook of Thick Film Technology', Electrochemical Publications, Ayr, Scotland (1976).
3   Snelling, E. C., 'Soft Ferrites', Iliffe.
4   Planar, G. V. and Phillips, L. S., 'Thick Film Circuits', Butterworth.
5   Leonida, G., 'Handbook of Printed Circuit Design, Manufacture, Components & Assembly', Electrochemical Publications, Ayr, Scotland (1981).
6   Dummer, G. W. A., 'Electronic Inventions and Discoveries', 3rd revised edition, Pergamon Press (1983).

Chapter 2

SUBSTRATES

J. F. PAWLING and G. H. SNASHALL

Mullard Mitcham, Mitcham, Surrey, England

2.1 INTRODUCTION

The choice of substrate for an electronic circuit rests largely on the ultimate use and specification of the circuit. In consumer applications cost is a major factor tending towards the use of the cheaper paper based PCBs. Professional equipment would be designed normally on FR-4 or similar glass epoxy materials. Where space is at a premium and temperatures are high, then hybrid technology in either thick or thin film is appropriate.

With leaded components and conventional mounting, the compatibility of the relative properties of substrate and component are not very critical. When considering surface mounting, the compatibility, particularly regarding expansion coefficient, becomes of vital importance. It must also be realised that circuits are seldom isothermal. While at first sight an alumina substrate and alumina chip carrier may appear to be perfectly matched, if the carrier contains a circuit dissipating power, then a considerable temperature differential may exist between the chip carrier and the substrate, thus stressing the attachment. Therefore the designer of a successful surface mounted assembly must be aware of the physical properties of the components, the substrate and the connection between the two.

This chapter will discuss the general requirements for surface mounting substrates and then describe the various types available and their properties.

2.2 GENERAL REQUIREMENTS FOR SURFACE MOUNTING SUBSTRATES

When considering a substrate for the surface mounted circuit, the following points must be first considered.

 (i) Compatibility with the components to be mounted.
(ii) Compatibility with the mounting machine.

2.2.1 Compatibility of Substrate and Components

As mentioned in the Introduction, this is principally a problem of thermal mis-match. As an example, we can consider the most commonly used component, a simple chip resistor, illustrated in Figure 2.1.

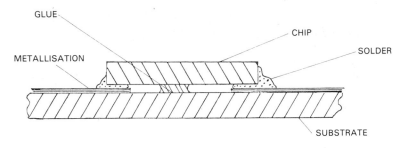

Fig. 2.1   A chip resistor mounted on a glass-epoxy substrate.

Chip resistors are made on an alumina substrate with an expansion coefficient of about 6 ppm/°C. Suppose this is mounted on a glass epoxy PCB with an expansion coefficient of about 15 ppm/°C. It is obvious that a mis-match occurs and it is necessary to examine the stresses induced in the system. The most important first step is to consider when the system is in zero stress. This is not as might be expected at normal room temperature, but is at the point where the solder solidifies, assuming it has been formed at new isothermal conditions. This, for normal 60/40 lead tin, would be about 190°C. Thus on cooling to room temperature the system will have the alumina in compression and the glass/epoxy in tension. The majority of the stress will appear as shear stress in the solder joint which may deform it. Providing all stresses are within the strength of the individual components, the assembly will be stable. Heating (isothermally) will reduce the stresses, cooling will increase the stress.

Taking this simple system, the stress can be reduced in 3 ways:

(i) Match the expansion coefficients of component and substrate.
(ii) Make the connection between component and substrate flexible so that its deformation absorbs the dimensional difference. This is the classic case of a leaded chip carrier correctly soldered, the leads providing the compliant link.
(iii) Make the substrate compliant, as in a flexible PCB, or use a special substrate with a flexible top layer.

The old engineering adage 'if it won't bend, it will break' is of particular significance in the problem of mounting systems with a thermal coefficient of expansion mis-match. Ideally a weak flexible link must be built into the system somewhere.

However, in practice one can usually get away with chip resistors, capacitors and small leadless chip carriers mounted even on glass epoxy PCB. This is because the dimensional differential is small enough to be absorbed by the relatively strong solder joint. Large leadless chip carriers are less satisfactory and can only be mounted if particular attention is paid to the size and flexibility of the soldered joint.

2.2.2 Compatibility with the Machine

Apart from the properties of the individual substrate and the method of laying down the conductors, three factors are vital and peculiar to substrates required for use in placement machines.
These are:

(i) An accurate and reproducible method of locating the substrate relative to the conductor pattern.
(ii) Uniformity of thickness.
(iii) Flatness unless the substrate is sufficiently flexible to have this parameter controlled by the rest of the machine.

2.2.2.1 LOCATING METHODS

The methods used for location will depend on the material of the substrate and the method of forming the conductor pattern. The two main methods are:

(i) Location on the edge of the substrate.
(ii) Mechanical location by holes or slots.

2.2.2.1.1 Location on the Edge

This is particularly applicable to substrates used in the thick film process. Alumina or similar materials are not easily machined or drilled accurately. Where conductors are printed (even more so if complex multilayered circuits are envisaged), a location in the printing nest is essential. While it is possible to locate a rectangular substrate on a corner, out-of-squareness can cause uncertainties. The best location method is to use three pins as in Figure 2.2.

Fig. 2.2 Locating a substrate using three pins.

When the substrate is pushed into the corner, even if corner A is not accurately at right angles, then the substrate can only locate in one position. The pin positions should be identical in the printing nest and placement machine nest, and some means used to identify corner 'A'. This may be just the asymmetry of the conductor pattern, or a definite witness mark may be employed. This edge location is virtually the only method possible for thick film substrates, as these are, in most cases, printed as multiple patterns. In order to maintain the location, the substrates must be presented to the placement machine as multiples and only separated after the components have been placed and attached.

2.2.2.1.2 Mechanical Location

This method is appropriate where the substrate material can be easily machined, punched or drilled, and is particularly suited to printed circuit boards. A very simple system is to have two holes, preferably of different diameters locating on pins in the placement machine nest. The position of these holes relative to the conductor pattern may be determined in several ways:

(i) The location is marked on the line film, and becomes an integral part of the conductor pattern, and is subsequently drilled.
(ii) If through-holes have to be formed, then the positions of the location holes can be determined by the drilling programme or incorporated in the punch tool. Once located, they may be accurately redrilled to size.
(iii) The location holes may be pre-formed and the line film accurately located relative to them before exposure of the resist.
(iv) The location features may be routed if the board has to be machined for any reason.

2.2.2.1.3 Location Accuracy

A generally accepted criterion of accuracy is that relative to the machine zero point, when located, all parts of the conductor pattern should be within $\pm 0 \cdot 05$ mm of their true position. In designing the location system, all parts and the nature of the substrate must be carefully considered to achieve the accuracy required by the placement machine.

2.2.2.2 UNIFORMITY OF THICKNESS

This is of particular importance in glued placement systems where the size and form of the glue dots are sensitive to the amount the dot is compressed. Not only is uniformity over the substrate required, but also from substrate to substrate. Where solder paste is used to attach the components prior to running the solder, the necessity for uniformity is somewhat less.

2.2.2.3 FLATNESS OR BOW

This again follows on from the necessity to place components accurately in height relative to the nest base. A very flexible substrate such as polyimide can be very easily accommodated by using vacuum to flatten it onto a nest base. For a more or less inflexible material such as an alumina substrate, the flatness must be inbuilt. It is desirable that the bow is not greater than $0 \cdot 03$ mm/cm.

Thin film substrates on ceramic pose a very difficult problem in location. Here the mask position has to be located relative to some transferable mechanical feature that can be picked up by the placing machine locator.

2.3 TYPES OF SUBSTRATES

The types of substrates which can be used for surface mounting technology can be categorised into 3 broad groups:

Ceramic Substrates
These are the familiar thick film substrates of alumina, beryllia and other ceramics including newer materials such as aluminium nitride.

Organic Substrates
These are copper clad materials based on organic resins reinforced in many cases by materials such as paper or glass fibre. They include all the familiar PCB materials as well as flexible polymers such as polyimide.

Inorganic Substrates
These describe the fairly recent group of substrate materials based on a metallic core coated with an inorganic insulator such as glass or glass ceramic.

2.3.1 Ceramic Substrates

These were the first type of material used for circuits that could be said to be 'surface mounted'. They have almost completely superseded glass which was the first material used to form thin film circuits. In these circuits the resistive and conductive elements are formed by deposition in vacuum. This is an expensive and time consuming process, although the conductors and other elements can be defined with great precision.

A cheaper system of laying down conductors is the thick film process developed in the '60s. Here the patterns are printed on the surface using materials which are essentially mixtures of conductive material and glass as an oxide loading agent. Figure 2.3(a) shows the conductor pattern and 2.3(b) the final assembly for a radio frequency thick film circuit. The conductors have to be fired at up to 900°C. For use with the thick film process, substrates are necessary having high melting points and stability at high temperatures. Many ceramics were experimented with but only two arc now commonly used. These are:

(i) Alumina
(ii) Beryllia.

Beryllia has a health hazard associated with it and is only used in cases where its extremely high thermal conductivity is essential. Two other materials in this group may be of interest in the future:

(i) Aluminium nitride
(ii) Silicon carbide/Beryllia mixture

2.3.1.1 ALUMINA SUBSTRATES

Alumina substrates as used in the electronics industry have settled down into two types:

(i) High purity alumina—99·5% aluminium oxide
(ii) 96% alumina.

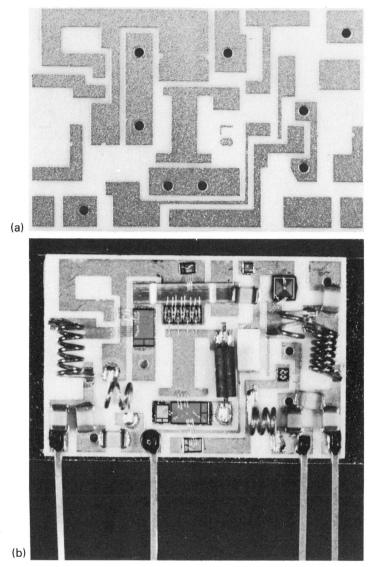

(a)

(b)

Fig. 2.3 The conductor pattern (a) and the completed assembly (b) for an r.f. application using thick film technology.

By far the most common is the 96% alumina substrate. Other percentages can be obtained but have little price advantage. The material consists of crystals of alumina embedded in a glassy matrix, usually calcium and magnesium silicates. At one stage, there were quite considerable differences in reaction with thick film conductor and resistor materials, dependent on the ceramic manufacturer. Nowadays the number of suppliers of alumina substrates for thick film work is quite limited, and little difference exists between the performance of ceramics of similar composition. The physical and electrical characteristics of alumina ceramics of different compositions are shown in Table 1: (Courtesy 3M Co. Technical Ceramic Products Division, Saint Paul, USA)

Table 1

	96% Al_2O_3 Alsimag 614	99·5% Al_2O_3 Alsimag 772
Specific Gravity	3·7	3·89
Safe Temperature-Continuous Heat (°C)	1550	1600
Hardness (Mohs)	9	9
Thermal Expansion Linear Coefficient (25-300°C)	$6·4 \times 10^{-6}$/°C	$6·6 \times 10^{-6}$/°C
Flexural Strength (kg/cm²)	3200	4900
Thermal Conductivity (W/m°K at 25°C)	35·1	36·7
Dielectric Strength (KV/mm)	8·3	8·7
Volume Resistivity —at 25°C	7×10^{14}	7×10^{14}
(ohm/cm cube) at 100°C	2×10^{13}	$7·3 \times 10^{13}$
at 300°C	$1·1 \times 10^{10}$	$8·6 \times 10^{11}$
Dielectric Constant —at 1 MHz, 25°C	9·3	9·9
at 10 GHz, 25°C	9·2	9·8
Loss Factor —at 1 MHz, 25°C	0·0028	0·0011
at 10 GHz, 25°C	0·0082	0·0004

In general, the much more expensive 99·6% alumina is only used in cases where its superior mechanical strength, or its excellent loss factor at high frequency, are required. It is used exclusively for microwave circuits where these features are essential. Such circuits are usually made using thin film methods, but with care conventional thick film materials can be used. Generally, however, their adhesion is less than on 96% alumina substrates.

2.3.1.2 BERYLLIA SUBSTRATES

Beryllia is a unique substrate material in that it has electrical and mechanical properties similar to those of alumina, but an extraordinarily high thermal conductivity—at 25°C it is higher than aluminium metal and eight times that of alumina. The properties of alumina and beryllia are compared in Table 2: (Extracts by Courtesy of Consolidated Beryllium Ltd, Milford Haven, UK)

Table 2

	99% Beryllia	99% Alumina
Bulk Density (g/cm²)	2·9	3·9
Hardness (Mohs)	9	9
Thermal Expansion (20-150°C) ($\times 10^{-6}$/°C)	4·5	5·5
Flexural Strength (kg/cm²)	1900	3100
Thermal Conductivity (W/m°K)	250	31
Dielectric Strength (KV/mm)	14	15
Volume Resistivity—at 20°C	10^{15}	10^{15}
(ohm/cm cube) · at 200°C	10^{15}	10^{14}
Dielectric Loss —at 100 MHz	0·0004	0·0003

Beryllia is used mainly for high power hybrid circuits. There are disadvantages, however:

(i) higher cost than alumina;
(ii) toxicity;

Substrates

(iii) lower mechanical strength;
(iv) thick film materials have been generally optimised for alumina.

In the case of fired substrates, provided adequate precautions are taken to control incoming material and the work area for dust, the hazard is minimal. References 1 and 2 give good accounts of the hazard and industrial hygiene necessary to use beryllia. It is unwise, without special precautions, to laser- or abrasive-trim components on beryllia or to use snapstrates. Most thick film materials are usable on beryllia. Some slight degradation of the adhesion of conductors is found and resistor compositions have to be characterised on the substrate, the ohms/square values being significantly different from those on 96% alumina. Once characterised they behave, and can be blended, in the regular manner.

2.3.1.3 ALUMINIUM NITRIDE

This is a very recently developed material, its main feature being good thermal conductivity. There is no perceived hazard as with beryllia. Properties are shown in Table 3: (Extracts by courtesy of W. C. Heraeus GmbH, Hanau, W. Germany)

Table 3

	Aluminium Nitride	Alumina	Beryllia
Density (g/cm²)	3·26	3·9	2·9
Hardness (Mohs)	7	9	8-9
Thermal Expansion (20-200°C) ($\times 10^{-6}$/°C)	5·6	5·5-8·5	6·5-6·8
Thermal Conductivity at Room Temperature (W/m°K)	40-170	10-35	150-250
Electrical Resistivity at Room Temperature (ohm cm cube)	10^5	$>10^{14}$	10^{13}-10^{15}
Modulus of Elasticity (KN/mm²)	300-310	300-380	300-355

Currently metallisation may be carried out using special materials developed by Heraeus. The material itself is grey/brown in colour and will be of interest to high power circuit designers in the future.

2.3.1.4 SILICON CARBIDE/BERYLLIA COMPOSITE

This is another contender in the good thermal conductivity stakes. As the material contains 20% beryllia, there is still a hazard associated with it. Properties are given in Table 4: (Extracts by courtesy of Hitachi Corporation, Tokyo, Japan)

Table 4

	Silicon Carbide/ Beryllia Composite	Beryllia	Alumina
Density (g/cm²)	3·2	2·9	3·8
Thermal Conductivity (W/m°K)	270	240	29
Volume Resistivity (ohm cm cube)	$>10^{13}$	$>10^{14}$	$>10^{14}$
Dielectric Constant—at 1 MHz	40	7	10
Dielectric Loss —at 1 MHz	0·05	0·0005	0·0002

It is claimed that metallisation can be carried out using regular thick film materials.

2.3.1.5 SIZE OF CERAMIC SUBSTRATES

The sizes of ceramic substrates obtainable are generally limited by the bow of the fired substrate. The most common size is 2 in. × 2 in., although 4 in. × 4 in. are easily obtainable and up to 7 in. × 7 in. are possible. Circuits delineated on ceramic substrates are generally small and many circuits can be formed at one time on a substrate. Circuits are then separated either by scribing between them and cracking the substrate or, less usually, sawing. Scribing can be by diamond or laser. Pre-scribed substrates—'snapstrates'—are obtainable with lines either laser scribed or pressed into the substrate in the green (prefinal fired) state.

When using surface mounting techniques it is usual to mount all components on the substrates as multiples before separating the individual circuits.

2.3.1.6 HOLES

Holes can be formed in substrates. Three methods are available:

(i) Drilling, using either diamond or ultrasonic.
(ii) Laser.
(iii) Formation on pressing.

Holes may be metallised by either thin or thick film means and this method of connecting front to back of the substrate is of particular interest where the circuit has an earthed backplane and isolated earthed areas are required on the front plane. A circuit using this technique is shown in Figure 2.4. For subsequent

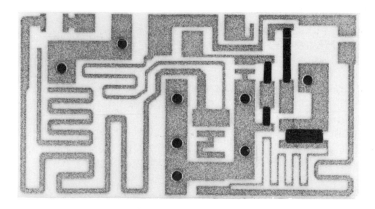

Fig. 2.4 Circuit using metallised holes to connect the earthed backplane to isolated earthed areas on the front plane.

metallising, holes drilled or pressed are easiest to handle. Sharp edges should be avoided where possible. Laser drilling is less satisfactory as, unless a spiral cut using a pulsed laser is used, truly round holes are difficult to form. Positional accuracy is good with holes formed by drilling or laser cutting. Where holes are pressed in the substrate, this has to be done at an early stage before final firing. Shrinkage on firing is considerable and very tight positional tolerances are not possible.

2.3.1.7   METALLISATION METHODS FOR CERAMIC SUBSTRATES

Hybrid circuits based on ceramic substrates were originally developed for the following reasons:

(i) Smaller dimensions could be obtained than with a PCB.
(ii) Resistive elements could be formed on the substrate.
(iii) Circuits could operate over a wider temperature range.
(iv) Mechanically very rugged.

With improvements of technique and the availability of surface mounting resistor elements, certainly (i) and (ii) are not as relevant today in the choice of substrate. Indeed once surface mounting by machine has been decided, the economics of forming resistive elements on the ceramic surface as opposed to placement need to be very critically examined in each case. Two methods of forming a conductor pattern on a ceramic are available:

(i) Thin film techniques.
(ii) Thick film techniques.

Historically thin film techniques were first because they could be successfully produced on glass substrates. Glass substrates are now completely obsolete, having been superseded by ceramics. A short account will be given of these two circuit forming systems.

2.3.1.7.1   Thin Film Technology

A good general account is given in Reference 3. The basic steps in the technique are as follows:

1 The substrate is cleaned externally and mounted in the evaporator.
2 Ion bombardment in the evaporator at about 1 torr is used for final cleaning of the surface.
3 A layer of nichrome is evaporated. This can be sufficiently thick to enable the resistive elements to be formed. Dependent on the circuit the ohms/square of the nichrome varies from 100 to 1000. The nichrome also acts as a bond coat and, where conductors alone are to be delineated, it need only be some nanometres thick.
4 A thin layer of copper is then evaporated over the whole surface.
5 The substrate is then covered with photo resist and exposed using regular photolithographic methods. After further washing, the resist is removed from the conductors while the open areas remain coated.
6 The substrate is then electroplated with gold or copper, or both, building up the conductor to the required thickness and surface finish.
7 The photo resist is removed and the thin layer of copper etched from the unwanted areas exposing the nichrome underlayers.

8 If resistors are not required, the nichrome layer is etched away in the open areas leaving the conductor pattern defined on the surface.

9 If resistors are required after stage 7, the substrate and conductors are coated again with photoresist and the resistors delineated by photolithography and etching.

Holes may be metallised through using the thin film process, but special evaporation methods are necessary to throw the metallising into the holes before plating.

The main advantages and limitations of Thin Film are as follows:

(i) Thin film gives excellent line definition to the limits of photolithographic equipment used. 50 μm (0·002 in.) line widths are easily attainable.

(ii) Thicknesses of metallisation depend to some extent on the photo resist used. The total plating thickness should equal the resist thickness. For good line definition resist should be thin.

As an example 100 μm (0·004 in.) lines up to 15 μm (0·0007 in.) thick are attainable with nearly rectangular cross-section.

(iii) It is essentially a one-layer process. Conductor crossovers are not possible. This obviously limits circuit layout freedom.

(iv) It is only possible to evaporate one thickness of resistor layer.

Therefore all resistors have the same 'ohms/square'. This can cause difficulties with resistor geometries. However, with the availability of individual surface mounting resistors and 'jumpers', some of the problems may be overcome by using placement techniques.

(v) The conductor resistances are metallic, thus very good performance at high frequencies is obtained. Indeed due to their cost, the main use of thin film technology is in the high frequency, low loss types of circuit requiring small and precise dimensions.

2.3.1.7.2   Thick Film Technology

This method of forming circuits on ceramic substrates is based on screen printing methods. The processes are described in detail in Reference 4, but a short account of the system is given here. Special materials have been developed based on glass containing compositions to form:

Conductors
Resistors
Dielectric or insulating layers.

These materials are made up into compositions satisfactory for printing by the addition of suitable vehicles and solvents. After printing, the organic materials are removed during the burn-out stage of the firing cycle and then the temperature is raised to cause the material to bond to the ceramic. Once fired, most materials can be refired several times. The process as originally developed requires firing in air to enable the organics to burn off and for the glassy materials to bond satisfactorily. One problem with this type of system is that conductor materials are confined to noble metals and their mixtures. However, successful systems based on copper and nickel as conductors have been developed.[5,6] These require firing in a nitrogen atmosphere using specially controlled furnaces. The copper conductor systems are very satisfactory. However, the associated resistor and dielectric are still only in a pilot experimental stage. Some of the apparent cost savings made by using base metal systems are however offset by the use of special furnaces and the cost of nitrogen gas. Most materials fire in the range 850-950°C and details are supplied by the individual manufacturers.

The noble metals—gold, silver, platinum and palladium—are the main materials used for the conductors in air fired systems. By far the most common is palladium silver. Two properties have to be noted by the user of thick film substrates:

(i) The noble metals are leached badly by the tin based solders normally used. Various mixtures—palladium silver and platinum palladium gold—have been used to attempt to minimise this effect. The effect of leaching is to produce brittle joints or breaks in the conductor. The manufacturer's recommendations should be strictly observed, and soldering temperatures and times carefully controlled. In extreme cases, say of soldering on gold conductors, special non-leaching solders based on Indium are available.[7]

(ii) The resistivity of the conductor systems, apart from gold or copper, is surprisingly high. For instance:

Palladium silver	= 30 mohm/square
Platinum palladium gold	= 70 mohm/square

whereas gold or copper are in the range 2-3 mohm/square.

Thus conductors carrying any appreciable current need to be adequately dimensioned to prevent unwanted volts drop. Conductors may be reliably printed down to about 100 μm (0·004 in.) with corresponding gaps. However, as shown in Figure 2.5, narrow conductors tend to have a higher resistance than their materials would suggest, due to the printed profile.

WIDE LINE   NARROW LINE

Fig. 2.5 Narrow conductors have a higher resistance than their material resistivity would suggest.

The same materials used for the production of Thick Film circuit resistors are also employed for the placed surface mounting components described in the next chapter. Most systems are based on ruthenium dioxide. Values of 'ohms/square' range from 1 to 10^8. The availability of this range of materials gives the possibility of producing resistors of the same physical size, but with an enormous range of values. Unless special precautions are taken, the printing accuracy, particularly of physically small resistors, is only ±20%. The usual scheme is to trim resistors which have been printed, aimed at a value 30% lower than the nominal. The resistor is cut using an abrasive nozzle or a laser shown in Figure 2.6. Using a probe system and a sophisticated laser trimmer up to 6 resistors per second can be trimmed to ±5% easily. There is an element of post trim drift of about 0·5%.

Being able to trim resistors this way enables a 'trim to function' system to be used, for instance, in an R-C oscillator. The circuit is operated and the resistor trimmed until the frequency is correct. This again, with suitable feedback, can be an operation requiring only a fraction of a second.

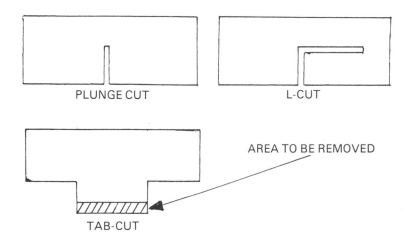

PLUNGE CUT   L-CUT

AREA TO BE REMOVED

TAB-CUT

Fig. 2.6 Thick film resistor adjusted by a laser.

However, the use of thick film resistors on a substrate must, if subsequent component placement is used, be affected by economics. If a large number of resistors on a substrate have the same ohms/square, then printing direct is probably economic. If there are only one or two, then placement is probably cheaper. The availability in quantity of surface mounting resistors has undermined the position of thick film techniques as the only way of producing compact resistors for use on ceramic substrates.

Insulating or dielectric layers can be printed using materials which are essentially insulating glasses usually of the crystallising variety. Insulation values of a typical 30 μm layer are of the order of 10^{12} ohms with breakdown in the region of 300 V. The main practical problem is the laying down of a void-free layer. Firing temperatures are in the range 700-920°C and most materials are designed to be capable of being re-fired with conductors.

2.3.2 Organic Substrates

These are the familiar PCB materials. The basic insulator layer is formed by an organic polymer and a suitable filler. While fillers can be selected from paper, glass, asbestos and cotton, and the resin from epoxy, phenolic, polyester and silicones, the two most common combinations are:

(i) Phenolic/paper.
(ii) Glass fibre/epoxy.

The filler is in the form of a mat or sheet and the resin is impregnated and processed in a press. In the more common subtractive process of producing the conductor circuit, a layer of copper foil is bonded onto one or both sides of the insulator. The circuit is then delineated by a photolithographic process. An alternative is to start with an unclad laminate and selectively deposit copper just where it is required by an additive process.

2.3.2.1 PROPERTIES OF PCB MATERIALS

The properties of a typical paper/phenolic and glass/epoxy insulator board are shown in Table 5 (Reference 8).

Table 5

Property	Phenolic/Paper	Epoxy/Glass	Units
Tensile Strength	100	300	MN/m^2
Flexural Strength	130	480	MN/m^2
Water Absorption	5	0·003	g/sq.m
Insulation Resistance	10^5	10^5	ohms
Permittivity at 1 MHz	5·0	5·3	-
Loss Tangent at 1 MHz	0·060	0·030	-
Expansion Coefficient	10	15	X10^{-6}/°C
Thermal Stability	fair	good	
Working Temperature	− 40 to + 100°	− 55 to + 140°	

From the point of view of surface mounting there is little difference in performance between the two, both having a high expansion coefficient relative to surface mount components. The choice will depend, not on the fact that the circuit is a surface mounted one, but on the circuit application.

Laminates for PCB manufacture can be obtained in a range of standard thicknesses. This dimension is expressed as a fraction of an inch such as 1/32, 3/64, 1/16, but these approximate to metric dimensions of 0·8, 1·2, 1·6 mm, and so on. The copper laminate can also be varied and is expressed in the weight in ounces per square foot. Commonly used thicknesses are ½ oz/ft^2 (nominally 0·0007 in.) and 1 oz/ft^2 (0·0014 in.). As the minimum conductor width that can be etched is a function of the copper foil thickness and the greater packing densities possible with surface mounted components will demand narrower conductors, laminates having thinner foils are likely to become more popular. Unlike the inorganic materials described earlier, most organic based substrates will burn if subjected to a naked flame or a continuous high voltage spark. For most applications flame retardancy is now a requirement and, to meet this need, alternatives to the more popular laminates are supplied in flame retardant versions. For example, FR-4 is the flame retardant version of G-10 and FR-5 the flame retardant version of G-11. These are all examples of glass cloth/epoxy laminates.

2.3.2.2 ADDITIVE PROCESSES

As mentioned earlier, most circuits are made by etching copper sheets bonded to the core material. This, for average circuits, means the removal of up to 80% of the copper with consequent etchant disposal problems. Circuits can be made by a process of sensitising the areas of substrate material to be covered with copper (tracks and pads) and then electroless plating to form the circuit.

Reference 9 gives a good account of the process. However, in spite of the environmental advantages, most circuit boards are made by the subtractive process.

2.3.2.3 CONDUCTOR DIMENSIONS

The width of conductors and the spacing between them are determined by a number of factors and include:

(i) The required component packing density.
(ii) The current carrying capacity and voltage isolation demanded by the circuit design.
(iii) Solder processing limitations such as solder bridging.
(iv) Cost effective manufacture of the substrate.

The use of Surface Mounted components assists in the achievement of greater packing densities. The current carrying capacity is a function of the resistance of the conductor. Using 1 oz/ft^2 copper the resistance is given by $R = 5 \times 10^{-4}$ ohms per square, e.g., a conductor 4 inches long and $0 \cdot 01$ in. wide has a resistance of $0 \cdot 2$ ohm. For earth and supply conductors, the resistance can be important, but in general is not a significant factor for the majority of signal lead conductors. The use of reflow soldering techniques enables closer conductor spacings to be handled (Chapter 7). Thus, the possible limitations of the PCB manufacturing process itself are likely to become increasingly significant.

The PCB copper pattern can be derived by an additive or subtractive process. With the subtractive method the pattern can be determined by screen printing, or by U-V radiation through a photographic negative or, more recently, by computer controlled laser scanning. Screen printing is the most economic, but the least accurate, the accuracy being dependent on the dimensions of the screen mesh, deformities of the screen and PCB-to-screen misregistration. The adhesion of the ink to the copper also tends to make the ink run beyond its defined limit line. The direct photographic image transfer methods using U-V light are more expensive but give greater accuracy. Using this latter method, the remaining main limitation in achieving minimum conductor width will be in the ability to control the etching process, particularly the undercutting which can occur and which increases with the thickness of the copper foil (typically $0 \cdot 002$ in. (50 microns) undercut with 1 oz/ft^2 foil). In practice, therefore, without resorting to thinner foils or additive processes, conductor widths of $0 \cdot 006$-$0 \cdot 008$ in. (150-200 microns) and similar spacings between parallel conductors are considered to be the normal limit. Thinner foils of 12 microns or less are available but involve greater problems in laminate manufacture.[10] An attractive alternative is the additive process briefly described in Section 2.3.2.2 with which 4 thou (100 microns) conductors have been produced in production quantities with acceptable yields.

2.3.2.4 OUTLINE SHAPING AND HOLE GENERATION

Organic material based boards, unlike ceramic substrates, are easy to cut to shape mechanically, and in the case of paper based materials, they can be punched, including the component attachment holes, although those intended to be plated-through generally need to be drilled. The necessity for holes apparently reduces when using surface mounted devices as they are not needed for component attachment and, although this is true for simple boards, the more complex circuits will require two or more interconnection layers and, when both sides of the substrate are used for component mounting, interconnection 'vias' or plated-through holes are needed to link up the circuitry on the various conductor layers. The main problem is then the surface area taken up by such a 'via' hole. The hole diameter must be sufficiently large to allow solder to flow through, where this is a requirement, and, furthermore, the minimum hole size that can be economically drilled is a function of the thickness of the laminate (typically one third).

2.3.2.5 MULTILAYER PCBS

Using laminating techniques, printed circuit boards can be made having many layers of conductor, and hence able to facilitate crossover connections. Typically between 4 and 12 layers are used, each layer being between 0·2 and 0·8 mm thick. Connections between layers can be made by metallising the walls of holes previously drilled between the conductor layers. These must be drilled, as mentioned earlier, as punched hole walls are considered too rough. Once the holes have been deburred and cleaned, copper is added by electroless copper plating, followed by electroplating to an adequate thickness. The greater component packing densities possible with surface mounted components are likely to increase the need for multilayer circuit boards in order to achieve the large number of conductor interconnects within a given area. On the other hand, sometimes, for the more simple circuits, the choice offered to the circuit layout designer between surface mounted or wire terminated components, and the possible use of zero ohm resistor jumpers, can enable a double-sided layout to be redesigned as a single-sided board or a four-layer to become a double-sided circuit (see Chapter 9.3).

2.3.2.6 CHIPSTRATES

This modification to the conventional PCB was developed by STC Exacta at Selkirk. The construction is shown in Figure 2.7. Essentially a flexible layer is

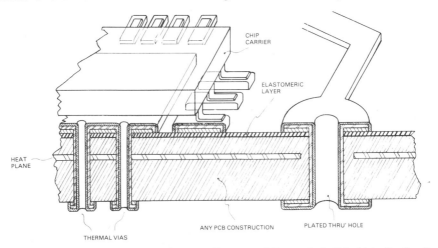

Fig. 2.7 The construction of a Chipstrate. (Courtesy of Exacta Ltd, Galashiels, Scotland)

bonded between the copper conductor and the rigid core material. This layer, after the copper circuit has been etched, acts as a flexible buffer and thus the difference in thermal expansion coefficients between the surface mounting components and the core layer is taken up. Chipstrate manufacture uses available printed wiring and related materials technology to produce fine line interconnection substrates suitable for surface as well as through mounting of components, packaged or not. As was mentioned earlier in this chapter, in practice for small components the expansion difference has little effect, but when mounting large leadless chip carriers, the presence of some flexibility is vital and Chipstrate elegantly supplies this.

2.3.2.7 FLEXIBLE CIRCUITS

These are now, in general, based on polyimide films with copper bonded to them. The circuit is formed by photolithography. Polyimide itself has a temperature range up to 400°C, but the adhesives used in bonding the copper limits soldering temperatures to 240-250°C. It is a very suitable material for use with high frequency circuits. Multilayers and plated-through hole techniques can also be employed. Surface mounted components can be mounted with confidence as expansion differences are taken up by the flexible nature of the substrate. However, because they are non-rigid, special attention has to be paid to the design of the nest on the placement machine. Vacuum hold-down on a flat plate is particularly effective. Some problems may arise in the provision of a sufficiently accurate locating system.

2.3.2.8 ADDITIONAL PROCESSES

Once the basic copper conductor pattern has been generated on the chosen laminate, a number of additional processes can usefully be incorporated. The attractiveness of copper as the conductor stems from its good bulk electrical conductivity, coupled with low cost. One disadvantage is its tendency to tarnish, thus reducing its solderability and its surface conductivity for plug and socket connections. The solution is to apply a further protective surface treatment to the copper while it remains untarnished. A common treatment in order to preserve solderability of the copper is to coat with an organic solderable lacquer. This is not recommended for use with surface mounted components due to the later curing and drying processes involved. A better solution is to coat with a tin or tin/lead alloy. Because of the small pad sizes the coating should be as flat as possible, otherwise components correctly positioned can slide out of place. For boards which are to be reflow soldered and carry only SMDs, electro tin plating provides a 6-8 μm thick finish which is at the same level as the solder resist. For mix print applications a hot air levelled tin/lead alloy finish of 15-20 μm is suitable for boards which are then going to be wave soldered.

Where the printed circuit design includes conductor edge connectors which are to be plugged into a socket, some form of plating to preserve good electrical contact is essential.[11] Inexpensive solutions are tin/lead or nickel, while gold directly on copper or gold on an intermediate undercoat of nickel is obviously more expensive, but often preferred.

The increased packing densities that can be achieved using surface mount components introduces greater possibilities for unwanted solder bridging and also increases the chance of voltage breakdown because of the reduced spacing between conductors. Both problems can be reduced by applying a solder masking coat to the non-conductor surfaces of the substrate.[12] For example, the minimum spacing between conductors can be reduced from 0·1 in. (2·5 mm) when uncoated to 0·06 in. (1·5 mm) for DC or AC peak voltages of 301-500 Volts. Other advantages of using a solder resist are to reduce the amount of solder consumed, reduce contamination of the solder in the reservoir and ease visual inspection of the soldered assembly.

Solder masks can be in liquid form which are generally screen printed onto the substrate and then cured using oven drying or exposure to ultra-violet radiation. Alternatively, a dry film resist can be laminated onto the substrate surface under vacuum. The film is photosensitive to U-V and hence the required pattern can be accurately defined by photographic means. Screen printing techniques are less accurate[13] and, for SMA applications where distances between pads or lines and pads are less than 0·015 in. (375 microns), dry film resists are recommended.

2.3.2.9 POLYMER CONDUCTOR RESISTOR MATERIALS

Conventional thick film processes cannot be applied to organic substrates due to the high firing temperatures. However, the attractions of the printing technology are very great in terms of material usage and production economy. While still in the early stages of development, systems akin to thick film conductor and resistor materials but using polymers instead of glass are available. The process then becomes identical to that described for ceramic substrates except that the 'firing' temperature is in the range of 150-250°C. Glass epoxy, high grade phenolic, polyester or polyimide substrates may be used.[14] When fully developed, the cost advantage of such a printed system, linked direct to a surface mounted component placement system, is very great. Some consumer products are already in the market place using such technology.

2.3.3 Metal-cored Substrates

There are two main types of substrate—those coated with glass, ceramic, or alumina designed to withstand the processes involved with the use of thick film inks and those coated with a polymer and aimed at replacing the organic printed circuit board. The main advantage over the latter is the much improved thermal conductance and ability therefore to dissipate heat. The prime advantage over alumina substrates is one of cost and the ability to be produced in various three-dimensional shapes, and to incorporate, if required, their own supporting brackets, flanges etc. The core metal can be selected from steel, copper, copper and Invar, iron and chromium and so on. A polymer/metal substrate (PMS) using an Invar-copper core structure[15] offers a further advantage compared with organic materials in that the thermal expansion coefficient can be matched to that of ceramic chip carriers. One of the disadvantages generally of metal-cored substrates is the large circuit capacitances involved. One proposal, 'Finstrate',[16] claims to overcome this by using PTFE, which has a low dielectric constant, as the insulator (Figure 2.8) on either side of a copper core. The softening temperature of the PTFE means that components must be mounted using a silver filled epoxy rather than soldered. One of the least expensive versions is a porcelainised steel which has been shown to dissipate heat better under some conditions than alumina. Suggested uses have included such cost conscious applications as heated food trays, room heaters, and front panels as well as telephone circuits and power supplies.[17]

One of the problems with metal-cored substrates has been the economic introduction of plated-through holes. The advent of SMA eliminates the need for holes except as 'via' connections between layers. It is likely that metal-cored substrates will become more popular in combination with surface mount devices for those applications where multilayers are not necessary, but power dissipations are high and production quantities are sufficient to pay for the high cost of initial tooling for blanking and punching of the substrate.

REFERENCES

1 Richardson, J., Hazardous Materials Bulletin No. 9, EEA, 8 Leicester Street, London, WC2 B7N.
2 Stokinges, H. E., Ed. 'Beryllium—Industrial Hygiene Aspects, Ed. H. E. Stokinges, Academic Press, New York.
3 Jones, R. D., 'Hybrid Circuit Design and Manufacture', ISHM, Marcel Dekker Inc.
4 Holmes, P. J. and Loasby, R. G., 'Handbook of Thick Film Technology', Electrochemical Publications Ltd, Ayr, Scotland.
5 Pitkanen, D. E., Cummings, J. P. and Speerschneider, C. J., 'Status of Copper Thick Film Hybrids', *Solid State Technology,* p. 141, October (1980).
6 Stein, S. J., Huang, C. and Cong, L., 'Base Metal Thick Film Conductors', *Solid State Technology,* p. 73, January (1981).
7 White, C. E. T. and Lepagnol, J. H., 'Special Solders for Micro-electronic Applications', Proceedings Internepcon, Brighton, England (1977).
8 Ross, W. McL., 'Modern Circuit Technology', Portcullis Press (1975).
9 Mansveld, J. F., 'The PD-R Additive Process and its Use in the Manufacture of Rigid Printed Circuit Boards', Proceedings of the First Printed Circuit World Convention, **Vol. 1,** June (1978).
10 'Know Your Laminates—Present and Future', BPA (UK) Ltd., Joint Report, *Electronic Production,* p. 49, October (1984).
11 Brown, A. F., 'Plating—Electroless and Electro', *Electronic Production,* p. 54, November/December (1982).
12 Manfield, H. G., 'Secondary Imaging with Dry Films', *Electronic Production,* p. 18, July (1983).
13 'Merits of Vacrel Dry Film Solder Masks vs Screen Printed Solder Masks', Brochure E1, Du Pont de Nemours & Co. (Inc.).
14 Bear, J., 'Polymer Thick Film Technology', *New Electronics,* p. 73, 15 October (1985).
15 Wright, R. W., 'Polymer/Metal Substrates for Surface-mounted Devices', Proceedings 2nd Annual Conference, IEPS, p. 445, November (1982).
16 Malhotra, A. K., Leabach, G. E., Straw, J. J. and Wagner, G. R., 'Finstrate: A New Concept in VLSI Packaging', *Hewlett-Packard Journal,* p. 24, August (1983).
17 'Porcelain Enamelled Metal Substrate Applications', Ferro Electronic VB Brochure, Netherlands.

Chapter 3

PASSIVE SURFACE MOUNTED DEVICES

G. A. WILLARD and E. G. EVANS*

Mullard Mitcham, Mitcham, Surrey, England
*EDS, Cheam, Surrey, England

3.1 INTRODUCTION

Modern component technologies, including Very Large Scale Integration, are far removed from early vacuum technology and the thermionic valves which made electronics possible at the start of the century. Throughout their development, active and passive components have always been mutually supportive. Valves needed from five to ten associated passive components; typical modern multi-function integrated circuits need about three. Despite this trend, passive components continue to outnumber active devices, and equipment assembly methods therefore need to accommodate both types.

Early electronic equipment used a chassis to carry valves, and small components needed long integral leads for attachment to the circuit. Later, when tag-boards came into use, components continued to need leads.

Today, with interconnections through the copper tracks of printed circuits, both axial and radial leads are still employed. Leads are bent, inserted into holes, soldered, and cropped. Some components have short, stiff radial leads for direct insertion. For automatic insertion, longer component leads permit packaging on reels; machines cut the excess before insertion, and generally 'clinch' the leads below the board to facilitate handling.

Surface Mounted Devices may be made by effectively deleting the wire lead attachment step in the components' manufacturing cycle. The Metal ELectrode Facebonded (MELF) device is made in this way. A MELF resistor is an axial leaded resistor with end caps which do not have leads attached to them. Connections to the device are by means of the solder-coated surfaces of the metal end caps of the component.

An alternative method of manufacture of a Surface Mounted Device uses a planar technique of deposition of materials on an inert base. SMD resistors are made by this method using a base of Alumina. SMD ceramic capacitors have a similar finished appearance but are made by building up successive layers of dielectric and conductor. This concept is not new; in special applications, some components of this type have proven capabilities, with a history that goes back to the early 1960s.

Components for surface mounting are now universally known by the name

'Surface Mounted Devices', or SMDs, although the name 'Surface Mounted Components' (SMCs) still appears occasionally.

These latter types of planar SMDs are generally referred to as true Surface Mounted Devices and are the more popular type of component with the majority of users. Surface Mounted Devices give many advantages to the assembly designer and manufacturer as a result of their small physical dimensions and ability to withstand flow soldering temperatures, thus enabling them to be mounted on the underside of the board. They also bring a number of new considerations:

(i) They need totally new kinds of packaging to allow them to be transported from the point of their manufacture to the point of assembly of the electronic equipment.

(ii) The substrates must be correctly designed to accept these components; as a consequence, design 'rules' are required to ensure that the substrate and the components are always fully compatible.

(iii) The size of the components creates a need for automatic assembly machines to remove the components from the packaging and to mount them speedily on the substrate.

(iv) Standard soldering methods must be modified to meet the requirements of these components, whether needing flow soldering of chips (by immersion in the solder wave) which have previously been attached to the substrate by means of adhesive, or reflow soldering of devices mounted in screen-printed solder creams.

(v) The quality standards and inspection methods must be modified to allow for the changed situation which the size and format of these components bring to assemblies which use them.

Considerable world-wide investment has already been made by electronic component manufacturers to offer the wide range of SMDs now available, a selection of which are shown in Figure 3.1. The assembly manufacturer needs to make only a relatively small investment to take full advantage of this assembly technique.

Fig. 3.1 A group of Surface Mounted Components or SMDs, indicating the general appearance of these small components. (Courtesy of Mullard Ltd)

3.2 SURFACE MOUNTED DEVICES—GENERAL CHARACTERISTICS

SMDs are small and have specially designed soldering contacts or, in some cases, short flat leads shaped for surface mounting. Because the design tends to preclude obvious leads protruding extensively beyond the general over-all outline of the component, SMDs are sometimes erroneously called 'leadless'. However, leads give mechanical flexibility to some larger SMDs where a mismatch in thermal coefficient of expansion between component and substrate could otherwise cause stress, possibly reducing component life in certain circumstances.

SMDs, both leadless and leaded, are constructed with certain well-defined characteristics: they are compact; they are directly solderable with no lead-bending or insertion in holes; they have solderable contact areas in a single plane, together with additional perpendicular 'wetting' areas to permit a good solder build-up, this being particularly desirable when flow soldering flat contacts; they also have carefully designed clearances below them as shown in Figure 3.2 to

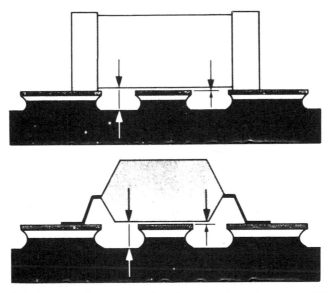

Fig. 3.2   Clearance for adhesive when flow soldering. (Courtesy of Philips)

facilitate initial attachment with adhesive if flow soldering is to be used; and they can withstand the high temperatures associated with soldering.

SMD dimensions are typified by the most frequently used (1206) SM resistor. This has dimensions of $3 \cdot 2 \times 1 \cdot 6 \times 0 \cdot 6$ mm, with an average clearance of 30 μm between the body and the plane of its contacts. Soldering areas contacting the substrate are $0 \cdot 35 \times 1 \cdot 6$ mm, and there are two perpendicular areas of $1 \cdot 6 \times 0 \cdot 6$ mm for solder wetting at the ends (these areas vary with component size). The 'footprint' or solder-pad dimensions for this '1206' resistor are typically squares of about $1 \cdot 4 \times 1 \cdot 4$ mm, separated by $2 \cdot 0$ mm. Other SMDs and their 'footprints' have similarly small dimensions.

If one considers the resistor to be the simplest surface mounted device, then at the opposite end of the range, there are the Surface Mounted Integrated Circuits. Typical dual-in-line versions (SO packages) may have 28 leads per side with a footprint consisting of $0 \cdot 5 \times 3 \cdot 1$ mm pads spaced at $0 \cdot 76$ mm intervals.

Alternative encapsulations include 'chip-carriers' and 'quad-packs' with leads on all four sides and a total lead-count in excess of 84.

The introductory remarks about SMDs given above apply generally to the whole range of devices, both active and passive. In this Chapter, details are given of passive SMDs including Surface Mounted resistors, ceramic multilayer capacitors, aluminium electrolytic capacitors, solid tantalum capacitors, 'chip-foils' and a variety of other more specialised components including connectors and sockets. Active components, including diodes, transistors, and integrated circuits are described separately in Chapter 4.

3.3   PACKAGING SURFACE MOUNTED DEVICES

Because packaging is such an important factor in the application of SM devices, details of component packaging are always given in the published data for each type of component. It is thus recognised that the user of SM devices will be aiming for full automation to maximise the advantages of Surface Mounted Assembly, and will demand packaging which adequately suits his machines. The packaging of these small components therefore takes on a high degree of significance.

There are six main ways in which these diminutive components are packaged: bulk, tube, rail, cartridge, magazine, and tape-on-reel. The choice of which packaging system is used depends largely on the requirements of the customer and is specified by him when defining his order to the manufacturer. However, the selection from which he can make this choice of packaging is determined by other factors such as availability, and suitability for his machine.

Bulk packaging in plastic bags or boxes demands that the assembly machine must have a hopper feed which includes a component-orienting system with, for some devices, an optical recognition system. Tubes and rails are methods of protecting larger components in transit and in storage; they may also be used for assuring the correct orientation of devices, but the packaging quantity is limited. Cartridge packaging is limited to small components which can move easily inside the cartridge. Magazines of various types and sizes are designed to fit specific machines, and carry only limited quantities of components. One or more of these packaging systems, however, will generally be offered by the component manufacturer along with tape on reel where this is possible.

Tape-on-reel packaging of SM devices (Figure 3.3) takes into account the main requirement that the devices should be able to be fed automatically and easily to assembly machines in large quantities. Orientation of devices within the tape is ensured by the component manufacturer, and this is preserved through transportation and storage; it is thus safeguarded until the tape cover is stripped open automatically inside the assembly machine for access to the devices in the tape.

Tapes can be fabricated in thin card, in which case they are generally known as 'paper tapes'. Alternatively, they can be constructed by embossing component-shaped compartments in plastic or plastic-coated aluminium, and these are then generally known as 'embossed tapes'. Aluminium tapes are preferred by the tape manufacturers because they are cold-formed whereas the plastic tapes need a heat process; furthermore, aluminium is more versatile when a wide range of 'pocket' sizes are needed for components with differing thicknesses and other dimensions. Tape widths are normally 8 mm, 12 mm, 16 mm, and 24 mm and the sprocket hole spacing is 4 mm. Component-compartments are included in the tapes at intervals of either 4 mm, 8 mm, or 12 mm (1, 2, or 3 sprocket holes) along the length of the tape. Wider tapes, up to 36 mm, are under consideration by some companies.

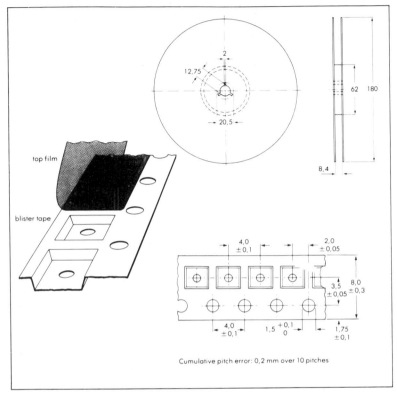

Fig. 3.3 Drawing of a typical embossed plastic tape or 'blister tape'. (Courtesy of Philips)

When the tape is filled for delivery, an empty lead-tape section is included for initial insertion into the assembly machine. The manufacturer will also specify a minimum filling-rate which ensures that almost every compartment is filled. The automatic assembly machine can detect omissions and can correct the assembly pattern to allow for these, but this slows down assembly slightly and is undesirable; filling rates approach 100% quite closely in practice.

Reels are similar to ciné-film reels. The diameter of the reel and the dimensions of the component compartment determine the maximum number of components that can be inserted in one reel.

Quantities packaged in any single tape can be small or large, the larger quantities simply employing a larger reel. The standard 8 mm tape on a 7 in. reel will hold up to about 5000 components of the 0805 or 1206 size; larger components naturally reduce the number that can be packed on a standard reel, but the larger reels permit several thousand larger SMDs to be packaged. Large reels also permit quantities of up to 10,000 small components to be packaged.

3.4   RESISTORS

The humble resistor sets many standards of size, quality and reliability for other Surface Mounted Devices. Perhaps unfortunately, however, its outward appearance is rather unimposing, and so it may deserve more than any other SMD the name 'chip' which is all too frequently applied loosely to SMDs collectively. Despite its lack of impressive external features, the SMD resistor is nevertheless the outcome of considerable ingenuity in design, and of relatively elaborate production procedures.

The most commonly-used style of SM resistor is the '1206' size; these digits give the two larger dimensions in hundredths of an inch (12×6), thus betraying the historical origins of these devices as being close to American space technology research centres. The final dimensions of the finished 1206 resistor are frequently given in millimetres as $3 \cdot 2 \times 1 \cdot 6 \times 0 \cdot 6$, but at the point when the user becomes interested in the precise measurements and their tolerances, each individual manufacturer's Data Sheets must be consulted because minor differences exist. A typical 1206 resistor is shown in Figure 3.4[1].

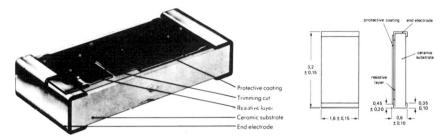

Fig. 3.4   See-through view of Surface Mounted resistor. (Courtesy of Philips)

The way in which good soldering is assisted by the design of the end-contacts can be seen in this diagram. The two coplanar flat surfaces at the ends touch down on to the copper on the substrate, assisting particularly in reflow soldering; in addition, solder can build up at each end by wetting up the 'vertical' faces, thus particularly assisting wave soldering.

The power rating of this 1206 resistor is classified as ¼ watt or ⅛ watt (depending on manufacturer) at ambient temperatures from $-40°C$ (or from some manufacturers, $-55°C$) up to and including 70°C; beyond this temperature the power is derated linearly to zero at 125°C.

There is also a less common $1/16$ watt SMD resistor (sometimes rated at ⅛ watt) known as the 0805, having nominal dimensions of $2 \cdot 0 \times 1 \cdot 25 \times 0 \cdot 5$ mm.

Within the single 1206 size, most manufacturers supply resistors in values in what are known as the E24, E12, and E6 series. These series have steps in nominal values and ±5% tolerance (E24), or ±10% tolerance (E12), or ±20% tolerance (E6) on the values themselves. For reference, the numerical values of the three series are given in Table 1. It should be borne in mind that the table of values does not *always* determine the actual tolerance of the resistor values themselves (although it generally does so). It is possible, for example, to obtain resistors with close tolerances such as ±1% and ±2% with values on the E96, E48, and E24 (±5%) ranges, although only precision circuit designs normally warrant the use of these resistors.

Table 1

Series of SMD Resistor Values

E24	10	11	12	13	15	16	18	20	22	24	27	30	33	36	39	43	47	51	56	62	68	75	82	91
E12	10		12		15		18		22		27		33		39		47		56		68		82	
E6	10				15				22				33				47				68			

Most suppliers provide values ranging from 1Ω to $10M\Omega$ in the E24 series, but the lowest and highest values in this range are sometimes supplied only in the E12 or E6 series. There is also normally a short-circuit 'zero-ohm jumper'. This 'jumper' is built with the same dimensions as the resistors in the range, has less than $50m\Omega$ resistance, and is normally rated at 2A. In the smaller 0805 style the values are generally between 10Ω and $1M\Omega$, also with a 'zero-ohm jumper'.

3.4.1 Manufacture of SM Resistors

In the construction of these diminutive resistors, a high grade alumina ceramic sheet forms the substrate. This starts off several centimetres square, and all the preliminary stages of manufacture for a large number of resistors are carried out on this sheet, more or less simultaneously for all of the individual resistors. The sheet is first scribed to provide fine 'crack' lines in two perpendicular directions; this ensures that individual resistors can be broken out of the sheet at an appropriate stage later during the manufacturing process. At the appropriate places, across one set of dividing cracks, internal electrodes are printed into position by a screen process on the upper face of the substrate sheet. The board is then baked to establish the effectiveness of these internal terminals. Following this, the resistive layer is similarly printed on to the board, and the board is baked again; the resistive layer which is deposited is always given a value of resistance lower than that actually required because the computer-controlled laser trimming process that follows can only raise the value. Laser trimming is applied to each resistor individually, in sequence, to trim up to the final value required; a small channel is burned out of the resistive substrate at considerable speed while test probes feed the resistance value to the computer which controls laser movement. Glazing is then applied by the deposition of suitable material over the central areas, excluding the internal terminals; this is then followed by another baking.

The primary substrate is then divided along all the scribed 'cracks' that lie in one direction to give long strips of partly constructed resistors, each resistor having both of its internal terminals facing one or other of the two newly exposed edges. At this stage, additional nickel intermediate end-electrodes are formed across the complete ends of the resistors, making good contact with the existing internal terminal at each end. A complete contact area now covers each end of each resistor. Final stages include the fracturing of the strips into individual resistors, and the plating-on of the external contacts which are designed for good soldering and for the conduction of current through the resistor. A diagrammatic representation of the various stages of manufacture[1] is given in Figure 3.5.

Materials used in the construction of these SM resistors include the following: alumina for the substrate, a paste based on ruthenium dioxide for the resistive element, silver (or gold if specially specified) for the internal electrode, nickel for the intermediate electrode, a tin-lead alloy for the external contact, and special glass for the protective coating. The nickel intermediate internal electrode provides protection for the first silver (or gold) electrode and prevents migration of the precious metal during manufacture, during soldering, and during operational life. (Note that the resistive material ruthenium dioxide is a material used by most manufacturers, but advantages may sometimes be claimed for alternative materials without declaring the specific ingredients.)

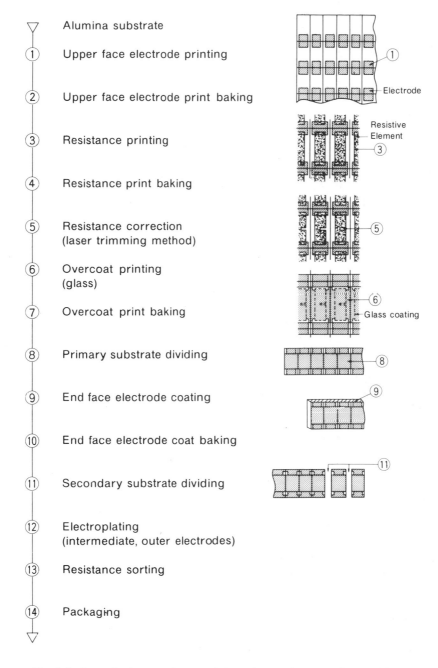

Alumina substrate
① Upper face electrode printing
② Upper face electrode print baking
③ Resistance printing
④ Resistance print baking
⑤ Resistance correction (laser trimming method)
⑥ Overcoat printing (glass)
⑦ Overcoat print baking
⑧ Primary substrate dividing
⑨ End face electrode coating
⑩ End face electrode coat baking
⑪ Secondary substrate dividing
⑫ Electroplating (intermediate, outer electrodes)
⑬ Resistance sorting
⑭ Packaging

Fig. 3.5 Stages in the manufacture of SM resistors. (Courtesy of Matsushita)

Marking of these minute SM resistors to indicate value and tolerance is not undertaken universally by all manufacturers for all of their ranges. The finished product is so small that marking is often in the form of a two-character or three-character code, although colour banding is also sometimes used. Typical codes are shown in Figure 3.6 (Courtesy of Matsushita), and the manufacturers of these

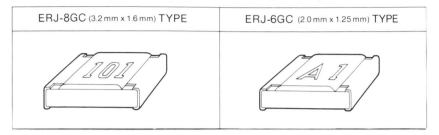

ERJ-8GC (3.2 mm x 1.6 mm) TYPE	ERJ-6GC (2.0 mm x 1.25 mm) TYPE

Fig. 3.6 Marking on some types of SM resistor. (Courtesy of Matsushita)

resistors offer marking on the reverse side if requested; the relevant published Data must be consulted for the interpretation of the code characters. In many practical situations, however, there is little need for such marking, since virtually all SM resistors are transported in the standard tapes and reels from the resistor manufacturer to the equipment manufacturer and to the assembly machine. The marking on the tape and spool is therefore normally sufficient.

3.4.2   Quality, Reliability and Performance of SM Resistors

Each resistor is checked for accurate resistance value at least twice during its manufacture. The manufacturing processes are also checked and controlled to ensure that the resistors are accurately dimensioned without any 'burrs' that might interfere with the intended automatic handling by assembly machines. Tests are performed at regular intervals to ensure that the resistors successfully withstand severe electrical and environmental conditions; some of these tests are standardised by such independent bodies as IEC and EIA, and others are self-imposed by the manufacturers. A few of these are mentioned below, but the manufacturer's literature or Data Sheets should be consulted for the details. From the figures given the SM resistor can be seen to be a truly tried and tested product.

3.4.2.1   ZERO-HOUR QUALITY AT DELIVERY

Acceptable Quality Levels (AQLs), to the IEC410 sampling system, require levels as low as 0·1% for electrical defects, and 0·65% for appearance or mechanical defects. Manufacturers can substantiate claims of meeting these requirements with significant margins.

3.4.2.2   RELIABILITY IN OPERATION

The typical Failure Rates quoted for standard SM resistors by some manufacturers and suppliers are $<1 \times 10^{-9}$ per component-hour at the standard 60% Confidence Level. This means a high probability of less than one failure for each one thousand million component-hours.

3.4.2.3 SOLDERABILITY

The solderability of standard SM resistors can be judged from typical figures given almost universally by manufacturers. Eutectic tin-lead solder is recommended. This has approximately 63% tin and 37% lead and will pass from the molten state to the solid state almost instantaneously on reaching the solidification temperature of approximately 185°C. Because the solder solidifies without going through a period of being semi-molten, there is virtually no tendency for the solder to leach out any of the metals used in the construction of the terminals of the resistors.

Under normal conditions, soldering can be achieved entirely successfully with the resistor immersed in solder at 230°C for a mere 2 seconds. Under unusual conditions, however, the resistor can withstand, if necessary, 60 seconds immersed in solder at 250°C, or 10 seconds at 260°C; these figures may not be confirmed by every manufacturer, and the reader is advised to check published Data Sheets for details before applying such figures to actual soldering situations. The usual maximum temperatures for reflow and wave soldering are shown in Figure 3.7, and it can be seen that the soldering capability of the resistor is well within the temperatures indicated.

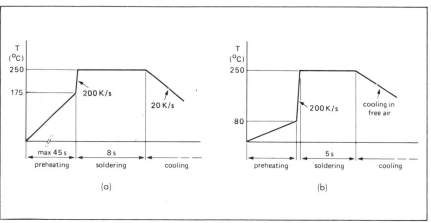

Fig. 3.7 Maximum temperatures which are usually encountered in (a) reflow soldering and, (b) wave soldering, both of which can be withstood by well-made SM resistors. (Courtesy of Philips)

3.4.2.4 TEMPERATURE COEFFICIENT

The SM resistor is a remarkably stable device under the conditions of varying temperature that are met in virtually any application. Generally, figures quoted for complete ranges of resistors indicate that all resistors in the range will change by no more than ±200 parts per million of the original value at the time of manufacture for each degree (Kelvin) change in temperature. This temperature coefficient (TC) is calculated as follows:

$$TC = \frac{R_2 - R_1}{R_1(t_2 - t_2)} \times 10^6 \text{ ppm/K}$$

where R_1 is the resistance value at the reference temperature t_1, and, R_2 is the resistance value at the test temperature t_2.

Some of the resistors, with extreme values of resistance either at the low end or at the high end of the range, may have somewhat poorer stability in changing temperature (a higher TC) than the majority of the resistors that occupy the main central portion of the range of resistance values. This is a detail that intending users of SM resistors may need to confirm with manufacturers. However, the maximum Temperature Coefficient represents the worst-case situation within the range, indicating that the majority of resistors have actual TC values well within the maximum limit figures quoted.

3.4.2.5 VOLTAGE COEFFICIENT

Stability of resistance value under varying applied voltages is another very desirable characteristic exhibited by these components. This characteristic is measured by applying a step-function change in voltage to the resistor, and measuring the instantaneous change in resistance before any slower consequential change in temperature can affect the resistance; the step in voltage is between the full Rated Continuous Working Voltage (RCWV) and 10% of that voltage. This gives a good indication of the resilience of the resistor under voltage changes over the major part of its working voltage range. The maximum acceptable Voltage Coefficient (VC) values for 'limit' components are generally quoted as 100 ppm/V, and the majority of resistors are well within this figure.

3.4.2.6 OTHER CHARACTERISTICS OF SM RESISTORS

There are many other features of the SM resistor which are the subject of many specifications and tests, and are also frequently mentioned in supportive literature describing the resistors. These cannot all be described here, but the user of these components may wish to obtain more information about these subjects from the manufacturers; there is generally no difficulty in finding the required data. Subjects that may be of further interest are as follows.

Short-time overload capability. For some resistors this can be 5 seconds at $2 \cdot 5 \times$ RCWV with only minute changes.

Voltage sustainable by exterior dielectric (glazing) cover. Frequently 200 V or, in some cases, 500 V.

Climatic categories. These are used in testing resistors. Example: IEC68, IEC115-X.

Humidity resilience. Also Load-life in high humidity.

Electrode strength. Affects handling procedures (and application to flexible substrate mountings).

Load-life. Also shelf-life.

Temperature-cycling capabilities.

Noise figures. Also the tests from which noise figures are derived.

Marking. Some manufacturers will mark resistors with special value codes.

This list is not exhaustive. However, the intending user of SM resistors will be able to examine the detailed information which all reputable component manufacturers are pleased to supply.

3.4.3 Cylindrical Leadless MELF Resistors

As an alternative to the 1206 and 0805 Surface Mounted rectangular resistors, the long-established cylindrical leadless resistors can be used in some applications. These resistors were designed primarily for the earliest generations of Surface Mounted Assembly machines and so some current machines which are designed to use rectangular SMDs are not able to handle them at all unless they are provided with an additional special 'pick-up' modification for use with these cylindrical devices. Other machines are designed specifically for handling cylindrical devices and are unable to handle rectangular SMDs.

Fig. 3.8   See-through view of MELF-outline SM resistor. (Courtesy of Matsushita)

Parts are as follows:
(1) alumina substrate;
(2) resistive element which may be carbon (Ni at values below 4·7 ohm) or NiCr or Ni depending on the resistor style, with laser-trimmed spiral cut;
(3) moisture resistant under-coat;
(4) heat-resistant body top-coat;
(5) plated steel end-cap;
(6) phenolic ink colour coding bands.

These cylindrical resistors (Figure 3.8) are still frequently used, however, because of their lower cost and because they have also frequently been stated to have higher dissipation ratings. Both of these apparent advantages have now been challenged by less expensive and up-rated rectangular SM resistors. The large-scale use of cylindrical SM resistors has been limited largely to the Far East, but the rectangular resistors have gained world-wide acceptance.

The cylindrical resistors are commonly known as MELF types, the letters standing for 'Metal-ELectrode Face-bonding'. Derived from the older wire-ended cylindrical resistor, these have metal end-caps for soldering, and they also have colour-coded value-bands. The ceramic core permits the formation of both carbon film and metal film resistors. Dimensions of the larger cylindrical resistor are 5·9 mm (length) and 2·2 mm (diameter). The smaller resistor is 3·45 mm × 1·35 mm.

One advantage which these cylindrical resistors have over the flat type is that they can be placed on the substrate without the need to ensure any particular vertical orientation; the flat type generally have to be placed one particular 'way up' and this is normally ensured by a previous correct location in the transportation tape. Since with the MELF type, this orientation is unnecessary, no tape is required and hopper feeding from bulk deliveries is possible; although hopper feeding is also possible with rectangular resistors, the necessary addition of an optical recognition system to control the feed usually militates against this procedure, and tape feed is seen to be preferable.

A second advantage of the MELF resistor is its excellent propensity for good soldering. The cylindrical shape of the end terminations provides an easily wetted surface for the solder to cover, and the interstitial curved wedge space permits the ingress and retention of sufficient 'working' solder to ensure a strong solder bond.

3.4.4 SM Resistors with Higher Power Dissipation Ratings

Power dissipation is limited to 250 mW for the standard 1206 resistor, and in some circuit designs this can be an inconvenient limitation. It has recently become possible to obtain 500 mW resistors which have dimensions $5 \cdot 2 \times 4 \cdot 0 \times 0 \cdot 7$ mm. These have values from 10Ω to $10M\Omega$ with tolerances down to 2%.

For even higher powers, the only current solution is either to use several standard chip resistors, or possibly the special resistors which have been modified from a leaded design to facilitate surface mounting. One such modified resistor (the Dale LVSR)[3] is available in 3 W and 5 W versions. Leads on the resistor are formed under the device, and special protrusions on the underside of the body moulding facilitate surface mounting.

3.4.5 Thick Film SM Resistor Networks

The Small Outline (SO) and Plastic Leaded Chip Carrier (PLCC) package configurations are used by Dale[3] to provide thick film resistor networks for surface mounting. The devices consist of a thick film network with edge pins attached, encapsulated in a rugged moulding with the pins formed into gull wing (SO) or J lead (PLCC) type. The SO packages are available in 14- and 16-pin versions and the PLCC packages in 20- and 24-pin versions. Resistor pairs, resistors with one lead common and separate isolated resistors are available in the packages. Figure 3.9 shows the 16-pin SO package version.

Fig. 3.9 Thick film Surface Mounted Resistor Network in 16-pin SO package. (Courtesy of Dale Electronics)

The highly stable thick film resistors have maximum individual power ratings of 150 mW and a total rating for the package of 750 mW, at 25°C. Available resistor values range from 100 ohm to 100 Kohm with tolerances of ±1, 5, 10 and 20%. Temperature coefficient is typically ±100 ppm/°C and the operating temperature range is −55 to +125°C.

3.5 SM CERAMIC MULTILAYER CAPACITORS

Ceramic Multilayer Capacitors, or CMCs, for Surface Mounted Assembly are similar to SM resistors in general appearance, and are packaged similarly; they are frequently called simply 'chip capacitors' but this name does not always adequately identity the type of capacitor.

As with resistors, the exterior of the capacitors gives very little indication of the expertise and skill needed in the design and fabrication of these diminutive components.

Ceramic Multilayer Capacitors take a wide range of body-sizes to accommodate the range of capacitance values required. The sizes required to achieve these capacitance values in fact determine the sizes for resistors and other small SM components, and the two smaller sizes of CMCs have dimensions identical to those of the common 1206 resistor and the less common 0805 resistor. The digits in the body-size number give the two larger dimensions of the body in hundredths of an inch; thus the two major external measurements of the 0805 body-size are 8 × 5 hundredths of an inch. Where dimensions are quoted in millimetres, they are selected from a range of values which have become an internationally accepted standard.

A 'see-through' view of a typical Surface Mounted CMC is shown in Figure 3.10 and some of the most frequently-used values of the dimensions, which are

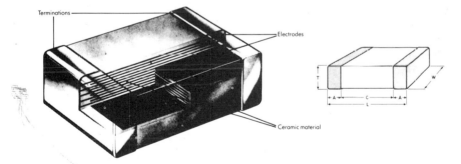

Fig. 3.10 See-through representation of Surface Mounted Ceramic Multilayer Capacitor. (Courtesy of Philips)

shown as L, W, T, A and C in Figure 3.10, are as given in the following non-exhaustive list of examples (there are smaller sizes, other intermediate sizes, and larger sizes). See Table 2.

Most suppliers provide standard 'off-the-shelf' values in the ±10% E12 series (see 'Resistors') from about 1 picofarad to in excess of 1 microfarad. Devices from 0·47 pF to 1·0 μF can be obtained direct (in bulk orders) from one major European manufacturer, for example, with considerably larger values currently planned. Tolerances different from the ±10% already mentioned can be specified if required; tolerances of ±1%, ±2%, ±5% and ±20% can be obtained from most

Table 2
Dimensions of Typical Body Sizes

Size Code	EIA-J No**	L (mm)	W (mm)	T* min. (mm)	T* max. (mm)	A* min. (mm)	A* max. (mm)	C min. (mm)
0805	732	2·0±0·15	1·25±0·15	0·51	1·27	0·25	0·75	0·4
1206	733	3·2±0·15	1·6 ±0·15	0·51	1·60	0·25	0·75	—
1210	734	3·2±0·2	2·5 ±0·2	0·51	1·90	0·3	1·0	—
1808	736	4·5±0·2	2·0 ±0·2	0·51	1·90	0·3	1·0	—
2220	738	5·7±0·2	5·0 ±0·2	0·51	1·90	0·3	1·0	—

* Both T and A vary with the capacitance value.
**Numbers of planned standard sizes approximating to sizes given.

manufacturers or suppliers. However, the tolerances depend largely on the kind of ceramic dielectric material used in the construction of the CMC; specific features of the various dielectrics control most of the facets of performance and have a major influence in determining costs. The dielectrics are described later.

3.5.1   Manufacture of SM Ceramic Multilayer Capacitors

One of the early stages of the construction of these capacitors consists of making the dielectric into an extremely thin film. Clearly, the thickness of this film necessarily has a major influence on the capacitance value. With the established manufacturing techniques, it is economical to make films having thicknesses of about 20 μm, and most commercial ranges of CMCs are constructed with dielectric of this thickness.

Research into the formation of thinner films indicates that routine production of films with thicknesses of about 2 μm may shortly be possible. Such films have the obvious attraction of offering an improvement factor of up to 10 in either the capacitance value (ten times larger) or in the size (one tenth of the volume). However, the film thickness is limited by the particle size of the ceramic polycrystal before the ceramic is made into sheets. Two factors conspire to make this an interesting and potentially rewarding research area which has already attracted considerable investment by the larger manufacturers. First, the mere formation of particles which are small enough to be made into such thin films represents a considerable task for a routine manufacturing process; second, it is necessary to reduce the statistical spread of particle sizes to an unusually narrow spectrum of values so that no unduly large particle can disrupt the consistency of the film. Thus, for a 2 μm thickness of film, the ceramic polycrystal particle diameter must be very much less than 1 μm with a narrow distribution of particle size. The resulting film must be without any flaw, without any significant perturbation in thickness, and without any perforation, however small. The fabrication methods which successfully meet these challenges economically at a full scale manufacturing level will herald the commercial arrival of the next generation of CMCs.

The best compromise at present is the film of about 20 μm thickness made generally from barium titanate. The required particle size is given by a solid-phase technique in which, initially, barium carbonate and titanium oxide react at about 1200°C, and then the dry, coarse powder which is produced is ground mechanically to reduce the size of particles. From the resulting material, reliable

20 μm film can be made. Alternative wet processes may soon become the source of thinner films, and later, vacuum deposition of ceramic material is expected to yield even thinner films.

In current capacitor production methods, the ceramic film is first checked to ensure that it is without internal voids, or 'pinhole' perforations. Strips of ceramic then have electrodes 'screened' into position, generally with a high-melting point metal such as platinum or palladium. Sections of film are then stacked with metal electrodes in two alternating positions ready for later connection to the two terminals (Figure 3.11) and the stacked sheets are formed

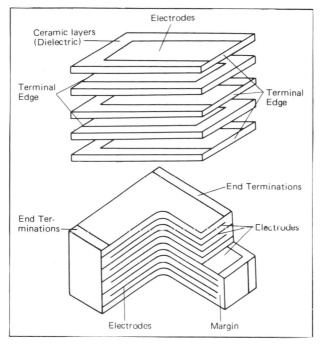

Fig. 3.11   Positions of ceramic layers and metallisation electrodes in relation to terminations in a CMC. (Courtesy of Kyocera) [Ref. 4]

into laminated units under high pressure. These laminated units are then cut into small blocks, each block representing the central core of one capacitor. After a 'binder burn-out' heating at a modest temperature, the blocks are fired at higher temperature to form virtually monolithic units: because this latter sintering process takes place at about 1300°C, the metals used for electrodes must have melting points well above that temperature. Platinum with a melting point of 1774°C and palladium with a melting point of 1552°C warrant the expense incurred because they meet the temperature requirements. When improved ceramics which permit a lower sintering temperature are brought from research and development into standard production, less expensive metals such as silver can be added to the electrode metals; this will reduce costs despite the additional requirement for including design features to inhibit internal silver migration. For the largest values of capacitance, nickel internal electrodes have been used; the reduction in material costs outweighs the disadvantages of the specialised fabrication techniques which have to be employed to avoid oxidation of the nickel.

Once the laminated blocks have been formed, it is next necessary to provide the electrical connections to the alternate conductive layers. Metal terminations are added at each end, again using a screening and firing technique. The metals which are used include silver and palladium. The proportions of these metals are selected to balance the several requirements as follows: it is essential to achieve reliable adhesion of the metal to the laminated block, such that the junction is capable of withstanding the later stages of fabrication, the subsequent soldering to a PCB, and also any other deleterious effects during use. Furthermore, resistance against leaching of the silver into the solder must be ensured, and the cost must be acceptable. When this terminal layer has been successfully added, layers of nickel and finally tin are plated onto the terminations; the nickel gives protection against deterioration through leaching, and the tin gives further general protection and also assists soldering. The 'see-through' view of a typical Surface Mounted Ceramic Multilayer Capacitor shown in Figure 3.10 gives an impression of the internal construction of the finished unit.

Alternative methods of layer-construction which are being assessed include, for example, procedures for forcing lead into minute air-gaps in ceramic layers; despite the obvious difficulties, the objectives of lower temperatures during construction, and lower material costs may be achievable.

An alternative type of ceramic capacitor is built in the MELF outline. This is described later in Section 3.6.

3.5.2   Dielectrics in Ceramic Capacitors

Although the resistivity of all of the available ceramic dielectrics is adequately high to ensure a high value of insulation resistance, the dielectrics differ in their loss angle and in the way that their capacitance and loss angle vary with voltage and temperature. Dielectrics fall into Classes and sub-groups of performance as defined by IEC and EIA; comparable alternative definitions from, for example, EIA-J and IJS, are virtually identical and can be cross-related directly by comparing details.

Only the most commonly used definitions are mentioned here because these represent practically all of the commercially available capacitors; the remaining definitions which would characterise a complete spectrum of possible types are, in practice, unused. The main Classes can be considered as follows.

Class I

These are low-loss ceramics which have high stabilities against temperature variation (minimal Temperature Coefficients), and near-zero ageing effects. Applications include resonant circuits, frequency filters, and timing circuits where capacitors with long-term stability are necessary.

Class II

These are high dielectric constant ceramics which have medium stabilities against temperature changes (medium Temperature Coefficients), and only limited ageing effects over long periods. Applications of the highest performance Class II capacitors include by-pass coupling and frequency discrimination circuits where the larger capacitances which are obtainable, together with their more than adequate stability, are desirable. Other Class II capacitors, with higher Temperature Coefficients, find application in non-critical coupling and de-coupling.

Class III

These are ceramics which provide maximum capacitance values in the standard body sizes, although this is accompanied by the highest Temperature Coefficients. They find application particularly in decoupling. (Class III capacitors are frequently fabricated in the MELF outline.)

In general, capacitors based on ceramics with the smallest temperature coefficients are the most expensive, and those capacitors intended for the less critical applications are proportionately less costly. Typical comparisons between Class I (NPO) and Class II (X7R and Z5U) are shown in Figure 3.12, but full details of all of the characteristics of particular ranges of capacitors based on any of the Classes of dielectric can be found in the published Data Sheets which are available from manufacturers or suppliers; the following points provide a means of comparison of the salient features.

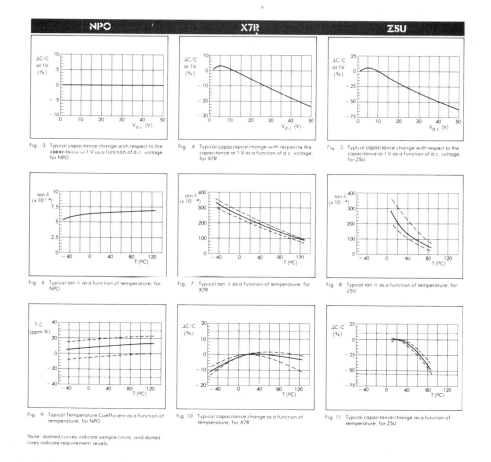

Fig. 3.12 Comparisons between selected types of CMC dielectrics. The left-hand column is for NPO (Class I), the central column for X7R (Class II), and the right-hand column for Z5U (also Class II). Top row is capacitance change with voltage, compared with capacitance at 1 volt. Second row is tan δ as a function of temperature. Bottom row for NPO is Temperature Coefficient as a function of temperature, and for X7R and Z5U is typical capacitance changes as a function of temperature. (Courtesy of Philips)

3.5.2.1   TEMPERATURE COEFFICIENTS

The Temperature Coefficient of one specific type of Class I ceramic is zero, that is, there is no appreciable change of capacitance over the relevant temperature range; the coefficients for other ceramics can be either positive (capacitance rising with increasing temperature) or negative (capacitance falling with rising temperature). Most Temperature Coefficients are negative on average for any particular ceramic, although the broadest tolerances on some Temperature Coefficients take account of some coefficients which are positive.

Temperature Coefficients with Class I capacitors are given in a way which indicates changes in ±ppm/K, that is plus-or-minus parts-per-million per Kelvin (degree Absolute). A code of letters (such as C = 0 ppm/K, P = − 150 ppm/K) used in the classification system is invariably given in detail in the published data.

Class II and Class III capacitors are frequently classified by a letter which indicates a percentage (such as B = ±10%, D = ±20%). These give the percentage by which the capacitance may change from the value at 20°C when heated to 80°C. The full scale is always given in published data.

Because most measurements for estimating Temperature Coefficient are made at 20°C and 80°C, this provides a satisfactory means of comparison of dielectrics and capacitors over most of their working temperature range. However, the real Temperature Coefficient may vary slightly from the value quoted when the capacitors are used outside the temperature range 20°C to 80°C, that is, when they are nearer to the limits of their practical operating temperature range.

Changes of capacitance value with temperature changes are not linear and so the Temperature Coefficient merely indicates the trend and the limits. When there is a requirement to be more specific about the temperature performance of capacitors, the manufacturer's published Data Sheet should be consulted. Most data sheets provide detailed graphs of such relationships, but it should be remembered that even these are statistical in nature.

3.5.2.2   TOLERANCES ON TEMPERATURE COEFFICIENT

Tolerances on Temperature Coefficients for Class I capacitors are indicated by a letter from a similar scale to that used for the coefficients themselves, but with a plus-or-minus significance (such as H = ±60 ppm/K, J = ±120 ppm/K). These ±ppm/K figures are plus-or-minus parts-per-million per Kelvin (degree Absolute) applied to the actual tolerance value itself.

Classes II and III capacitors have their tolerances on their Temperature Coefficients given by a letter which indicates a percentage (such as V = ±7·5%, X = ±15%). These are expressed as a tolerance on the actual capacitance value itself, and should be added to the Temperature Coefficient which is expressed as a percentage of the capacitance.

3.5.2.3   OPERATING TEMPERATURE RANGES

The operational temperature range of capacitors is defined by the manufacturer in such a way that adequate longevity for the component can be ensured. Despite this 'arbitrary' aspect, wide temperature ranges such as − 55°C to + 125°C, or − 25°C to 85°C are offered depending on the type of capacitor.

It should be noted that these temperature limits all exceed the narrower temperature range + 20°C to 80°C which characterises the value of Temperature

Coefficient (TC). The TC may therefore deviate from the quoted value when capacitors are used near these operational temperature limits which are given above.

3.5.2.4 DIELECTRIC STRENGTH

The strength of the dielectric is sufficient to withstand reasonable temporary over-voltages. These are normally given as 2·5 times greater than the rated voltage of the capacitor, for a period of two seconds, with the step-function charging current limited to 50mA.

3.5.2.5 INSULATION RESISTANCE

The insulation resistance is normally measured at the rated voltage, and is given both for 25°C and for 125°C. Each range of capacitors will have its own characteristic resistance value (see the manufacturer's Published Data), but the following values are frequently stated for Class I and some Class II capacitors. At 25°C insulation resistance is 100 GΩ (that is, 100,000 MΩ) or alternatively 1000 ΩF, whichever is the smaller. At 125°C these figures are reduced by a factor of ten to 10 GΩ and 100 ΩF.

3.5.2.6 DIELECTRIC LOSSES

Dielectric losses for Class I CMCs are sometimes quoted in terms of the Q-factor. Typical values given are $Q \geqslant 400 + (20 \times C)$ for values from 1 to 30 pF and $Q \geqslant 1000$ for values above 30 pF where C is the nominal value of capacitance in picofarads. Measurements are made at 1 MHz±0·1 MHz.

Alternatively, the tangent of the loss angle may be quoted, a typical value being Tan $\delta \leqslant 0 \cdot 001$ for C>50 pF, with low-value capacitors having an additional small capacitance-related factor included.

Dielectric losses for Class II and Class III capacitors are normally quoted in terms '042of the tangent of the loss angle, and are generally greater than for the Class I types. Measurements for Classes II and III are made at 1 kHz±0·1 kHz.

3.5.2.7 AGEING

Almost all ceramic capacitors lose some of their original capacitance value with age, the loss following a logarithmic law. Such losses are small and vary from one type of ceramic to another. Where such ageing losses might be considered important, the manufacturer's data must be consulted. In general, Class I capacitors will age less than Class II which, in turn, will age less than Class III. The most stable ceramics in Class I, frequently identified as COG types or an equivalent, do not in fact age by any significant amount over any reasonable projected life period, but these are exceptions to the general rule.

Ageing losses are normally quoted as a percentage loss related to a previous capacitance value, the loss occurring over a time 'decade'. Class I capacitors have very low ageing constants, and Class II capacitors will generally have ageing constants of between 1% and 5%. If the manufacturer quotes the original capacitance value as being within the selected tolerance at 1000 hours (as is frequently the case), then subsequent deteriorations should be predicted by assuming that the percentage loss quoted will occur in the next decade to 10,000 hours, and so on. Thus, a capacitor with about 1% per decade ageing constant will have lost a total of about 2% of its capacitance after one year and about 3% total after ten years, indicating a good level of overall stability.

The ageing constant is for a particular quoted ambient temperature. Operating the capacitors at elevated temperatures reduces the rate of ageing. (Raising the temperature artifically beyond the Curie temperature restores all of the original capacitance which then starts to age again normally at reduced temperatures.)

3.5.2.8 VOLTAGE RATINGS

In each of the different types of ceramic material, CMCs are normally obtainable rated at 50 V (EIA) or 63 V(IEC). Some manufacturers also supply 25 V and 16 V versions.

3.5.3 Quality and Reliability of CMCs

CMCs are rugged and yield good reliability and life measurement figures despite the rigours of SMA soldering techniques; all manufacturers state clearly that their capacitors can withstand the high temperatures associated with reflow soldering and the thermal shock associated with high-speed wave (immersion) soldering and, typically, immersion in solder at 260°C for up to ten seconds causes no damage.

Quality and reliability levels for individual ranges should be judged from figures which are given in manufacturer's Data Sheets. In general, the figures for CMCs indicate comparability with figures given for resistors.

3.5.4 High Value and High Voltage Ceramic Capacitors

SM capacitors with unusually high capacitance values and voltage ratings are available from a limited number of manufacturers including AVX[5], Novacap[6], Semtech[7] and Sprague[8]. Large capacitance values in the 60D8 size (0·60 × 1·40 in.) include 10 μF rated at 500 Volts, and 82 μF at 50 Volts. High voltage capacitors in the 6560 size (0·65 × 0·60 in.) include 390nF rated at 1000 Volts and 12nF at 5000 Volts. Novacap also offer a 6000 Volt capacitor range with values up to 3·9nF in the 5248 size. Specials with ratings as high as 10,000 Volts are offered by AVX.

3.5.5 Mica Chip Capacitors

A range of Mica Chip Capacitors is made by Wimpey-Dubilier[9]. These are available in sizes of 2·0 × 1·25 × 1·4 mm to 5·6 × 5·0 × 2·0 mm covering the range 1pF to 2000pF at 100 Volts and 1pF to 1200pF at 500 Volts. The devices offer very close tolerances down to ±0·25% and temperature coefficients of 50 ppm/°C.

3.6 CYLINDRICAL MELF CERAMIC CAPACITORS

The MELF (Metal ELectrode Face-bonding) ceramic capacitors are cylindrical and therefore do not need 'vertical' orientation perpendicularly to the PCB during placement. Transportation tapes are thus not essential, and so hopper-feed from bulk can be employed with some SMA machines.

The capacitance values and voltage ranges available with MELF-type ceramic capacitors are similar to those for rectangular CMCs, but there are many more Class III types. The entire spiral edge of each metallisation in the cylindrical winding makes electrical contact with its own termination, and so the capacitor has low inductance despite the spiral winding.

Typical dimensions of MELF ceramic capacitors are $5 \cdot 9 \pm 0 \cdot 2$ mm (length) and $2 \cdot 2 \pm 0 \cdot 1$ mm (diameter). The diameter of the MELF soldering terminal is generally slightly smaller than that of the body of the capacitor. As the body itself generally has a 'waist' with a slightly narrower diameter at the centre, real or dummy copper tracks should run under the capacitor to help ensure good attachment if adhesive is used; wave soldering then helps to ensure the bridging of any small residual gap which may exist between the MELF soldering-cap and the copper 'footprint'.

3.7 METALLISED POLYESTER SM CAPACITORS

Frequently known collectively as 'chip-foil' capacitors, the various types of metallised polyester capacitors are generally fabricated from sheets of flexible dielectric such as polyethylene terephthalate carrying a vacuum-deposited aluminium electrode. The metallised layers are interleaved with their alternate edge-contact areas on opposite sides and are then wound into a flattened cylindrical shape. The entire length of the outside edge of each metallisation is in contact with its own external contacts. This economic design yields high capacitance values with low internal inductance.

Encapsulation dimensions vary from $5 \cdot 5 \times 4 \cdot 2 \times 2 \cdot 3$ mm to $7 \cdot 3 \times 7 \cdot 0 \times 4 \cdot 0$ mm (all dimensions $\pm 0 \cdot 2$ mm), giving values from $1 \cdot 0$ nF to $0 \cdot 47 \, \mu F$. The encapsulation around the flattened cylindrical winding is a flame-retardant epoxy resin and this allows the capacitor to be subjected to the temperature-time stresses of either reflow soldering or immersion in molten solder during wave-soldering.

Typical performance of these Surface Mounted Polyester Capacitors is similar to that of their normal leaded counterparts with a voltage rating of 50 Volts DC and self-inductance of 5 nH. They are available in bulk packaging or taped and reeled in 12 mm wide embossed tape. (See Figure 3.13).

Fig. 3.13 Metallised Polyester Surface Mounted Capacitors. (Courtesy of Siemens)

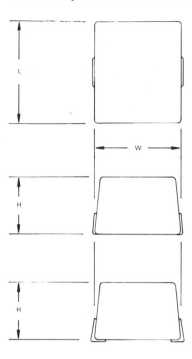

Fig. 3.14   Outline drawing of a typical metallised polyester SM capacitor. (Courtesy of Philips)

There are two types of terminal (see Figure 3.14). In both types short flat 'leads' emerge from opposite sides of the capacitor; in some designs these are across the shorter plan-view 'width' dimension, and in other designs across the longer 'length' dimension. One type of terminal extends only to a point where it is flush or level with the face of the body of the capacitor which will be adjacent to the PCB after assembly, there being no metal extending below the body of the capacitor. This type is best suited to wave soldering because the close proximity of the capacitor body to the PCB gives good adhesion before soldering, and the solder is able to wet the side contact and form the standard solder fillet. The alternative design has 'wrapped-under' extensions to the terminal, and this is more effective with either reflow soldering or epoxy bonding, where face-to-face contact between the terminal metal and the PCB copper 'footprint' is desirable.

3.8   ALUMINIUM ELECTROLYTIC SM CAPACITORS

Only a very limited number of manufacturers provide SM aluminium electrolytic capacitors. Philips of The Netherlands provides a full range from $0 \cdot 1$ μF to 22 μF (E6 values), with larger values planned. Voltage ratings are from 63 to $6 \cdot 3$ V. Two case-sizes, called 'case-size 1' and 'case-size 1a', are used, and an impression of the appearance of these capacitors can be gained from Figure 3.15.

These capacitors are constructed by first rolling aluminium foil electrodes with paper separators and then encapsulating the rolled unit with electrolyte in an aluminium cylinder. This cylinder is then inserted into a moulded outer plastic case. Connections are formed at each end, and the contact 'feet' for soldering are

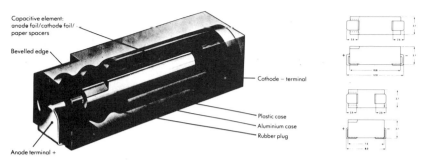

Fig. 3.15 See-through view of aluminium wet-electrolytic SM capacitor. (Courtesy of Philips)

wrapped under the moulding, leaving clearance for additional copper tracks under the capacitor, if required. The plastic used for the moulded exterior is capable of withstanding up to 10 seconds immersion in solder at 260°C, permitting soldering by normal SMA wave or reflow methods. Values of capacitance, and voltages, are marked with a simple code on the exterior plastic moulding of the capacitors.

Transportation packaging for completely automated handling is in plastic blister tapes, but boxes of loose capacitors can also be supplied.

Performance of these exceedingly small electrolytic capacitors is comparable in every way with that of the conventional versions which are built in a similar way.

A similar range of devices from Nippon Chemi-Con[11] covers 0·1 μF to 22 μF at voltages from 50 to 6·3 V. Two case sizes A and B are used with dimensions 6·4 × 4·6 × 2·5 mm and 7·4 × 5·2 × 4·5 mm.

A range of solid electrolyte aluminium capacitors for surface mounting are available from Philips Electronics. The 126 series offer a nominal capacitance range of 0·1 to 68 μF at voltages of 40 to 6·3 volts. Tolerance on capacitance is ±20% or ±10% and the operating temperature range is − 55 to + 125°C extending to + 175°C with voltage derating. The devices are suitable for all current soldering methods and heat resistance is specified as 10 seconds at 260°C. The capacitors are in fact SMD versions of the Philips 122 series with electrical parameters identical to these devices. The 126 series are rectangular capacitors which are encapsulated in blue cases, 6·5 mm long, closed at both ends by soldered copper caps. All of the five case sizes fit into 12 mm blister tape.

3.9 SM SOLID TANTALUM CAPACITORS

Surface Mounted solid tantalum capacitors have earned an excellent reputation over many years. These capacitors are noted for high capacitance-voltage product per unit volume and for high reliability. High costs associated with early versions of these capacitors have been much reduced by modern fabrication techniques, and these tantalum electrolytic capacitors are now more widely used.

Each capacitor contains a near-rectangular anode of high-purity sintered tantalum with an electrolytically-formed oxide dielectric layer. This anode is encapsulated with its solid electrolyte in an extremely rugged body, to which shock-proof and vibration-proof terminals are attached. (The structure enables the capacitors to operate satisfactorily at extremely high or low atmospheric pressure, far beyond the limits for wet types.) One manufacturer provides two alternative encapsulations, either with small 'wrap-under' contacts which are ideal for automatic placement machines and standard SM soldering, or alternatively with extended tab contacts which facilitate welding as well as soldering.

Solid tantalum capacitors are highly efficient in their use of board area and total volume. Area efficiency can be as high as approximately $8 \cdot 6$ μFV/mm^2 (5550 μFV/in^2), and volumetric efficiency as high as $6 \cdot 1$ μFV/mm^3 (100,000 μFV/in^3), ranking with the best large wet electrolytic capacitors.

Several manufacturers supply a range in E6 values from $0 \cdot 10$ μF to 100 μF, with tolerance values of $\pm5\%$, $\pm10\%$ or $\pm20\%$, and with rated voltages from 50 to 4 V. Generally, eight different case-sizes are used, most manufacturers keeping to a strictly rectangular shape, while others use a somewhat 'rounded-off' rectangular shape of a similar size. A typical rectangular type is shown in Figure 3.16, and the dimensions are shown in Table 3.

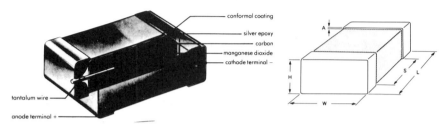

Fig. 3.16 See-through view of a typical tantalum solid-electrolytic capacitor. (Courtesy of Philips)

Table 3

Dimensions of Case-Sizes of Tantalum SM Capacitors

Case Size *	Width(W) ±0·04 (mm)	Length(L) ±0·04 (mm)	Height(H) ±0·04 (mm)	Spacing (S) (mm)	Clearance (A) (mm)	Clip-width (B) (mm)
a	1·27	2·54	1·27	1·02	0·127	0·762
b	1·27	3·81	1·27	2·29	0·127	0·762
c	1·27	5·08	1·27	3·55	0·127	0·762
d	2·54	3·81	1·27	2·29	0·127	0·762
e	2·54	5·08	1·27	3·55	0·127	0·762
f	3·43	5·59	1·78	4·06	0·127	0·762
g	2·79	6·73	2·79	4·19	0·127	1·290
h	3·81	7·25	2·79	4·70	0·127	1·290

*Case-sizes are given different code letters by various manufacturers in place of the a-to-h used here.

Table 4 gives the capacitance obtainable at each voltage rating for the case sizes which were defined for Figure 3.16.

Various manufacturers offer ranges of capacitors similar to those specified in Table 4, some with values down to $0 \cdot 01$ μ and with voltages down to 2 V. Maximum case size is $8 \cdot 0 \times 4 \cdot 6 \times 5 \cdot 0$ mm.

High temperatures associated with reflow and wave (immersion) soldering are well within the capabilities of these capacitors, some of which can tolerate 300°C under certain conditions. However, they may be susceptible to damage from thermal shock, and to ensure that the full potential reliability is realised in the application, a 'slow' pre-heating of both the board and the tantalum capacitors is

Table 4

Capacitances and Voltages in various Case-Sizes* with Tantalum Capacitors

Cap µF	4 V	6 V	10 V	15 V	20 V	25 V	35 V	50 V
0·10	a	a	a	a	a	a	a	a
0·15	a	a	a	a	a	a	a	a
0·22	a	a	a	a	a	a	a	b
0·33	a	a	a	a	a	a	b	b
0·47	a	a	a	a	a	b	b	c
0·68	a	a	a	a	b	b	c	d
1·0	a	a	a	b	b	c	d	e
1·5	a	a	b	b	c	d	e	f
2·2	a	b	b	c	d	e	f	f
3·3	b	b	c	d	e	f	f	g
4·7	b	c	d	e	f	f	g	h
6·8	c	d	e	f	f	g	h	-
10·0	d	e	f	f	g	g	-	-
15·0	e	f	f	g	g	h	-	-
22·0	f	f	g	g	h	-	-	-
33·0	f	g	g	h	-	-	-	-
47·0	g	g	h	-	-	-	-	-
68·0	g	h	-	-	-	-	-	-
100·0	h	-	-	-	-	-	-	-

*Case-sizes are identified with different letters by various manufacturers.

frequently recommended. Optional gold-plated terminations are available if required.

These devices were originally developed for hybrid circuits in military applications. Most manufacturers supply these relatively expensive devices in small quantities only: standard blister packs contain 50 pieces, and bulk packages contain 1000. New and more cost-effective ranges, designed for industrial and consumer applications, are constantly being introduced with comparable technical specification and reliability. These are available in bulk or in tape and reel.

3.10   MINIATURE SM INDUCTORS

Wire-wound inductors which are specially designed for automatic assembly can be added to an SM substrate. Although such inductors are made by only a limited number of manufacturers, an extremely comprehensive range of values is nevertheless obtainable. Some typical types are shown in Figure 3.17.

A range of inductance values from 1·0 µH to 390 µH can be obtained from one manufacturer (muRata) in a single-size design having external dimensions 3·2 × 2·5 × 2·0 mm   (length × width × height).   Current-carrying   capability   is 100 mA at the low end of the inductance range, and 15 mA at the high end. A complementary range from the same manufacturer has inductance values from 1·0 µH to 470 µH in a somewhat larger design (5·0 × 2·5 × 3·15 mm), with current-carrying capability from 300 mA to 80 mA.

For this type of inductor the Q-factor is between 35 and 50; the resistance is

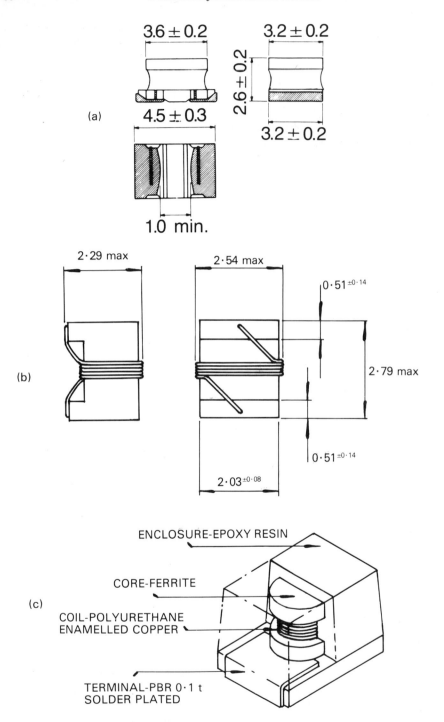

Fig. 3.17 Typical SM inductors which are capable of withstanding SMA soldering and washing techniques. (a) Courtesy of muRata [Ref. 13], (b) Courtesy of Coilcraft [Ref. 13], (c) Courtesy of Matsushita)

dependent upon the inductance selected because this determines the choice of wire. A brief selection from a typical range (muRata) is shown in Table 5, but the full Published Data should be examined to provide details of all available values; this selection gives some indication of the availability of 'off-the-shelf' SM inductors.

Table 5

muRata Part Number	Nominal Induct. (µH)	Tolerance (±%)	Q min.	DC Resistance max(Ω)	Min. Self-res. Freq. (MHz)	Allowable Current max(mA)
LQN3N1R0M	1·0	20	35	0·25	120	100
LQN3N8R2M	10	10	40	1·3	20	50
LQN3N101K	100	10	50	7·0	8·0	20
LQN3N391K	390	10	50	15·0	4·5	15
LQN5N1R0M	1·0	20	35	0·2	120	300
LQN5N100M	10	20	40	0·6	23	270
LQN5N101K	100	10	40	2·9	8·0	150
LQN5N471K	470	10	40	10·0	3·0	80

Note: Values are only a selection from the full range.

Both of these muRata ranges, which are quoted as examples of SM inductors, are capable of being delivered in the standard 8 mm tape for automatic assembly machines. Because the larger inductors are relatively tall, the tape compartments have to be deep; the length also necessitates placing these components at 8 mm-pitch positions in the tape (which still has 4 mm-pitch sprocket perforations) and the standard 7-inch reel then carries only 500 inductors; the smaller inductors permit packaging of 2000 in smaller compartments at 4 mm-pitch positions in tape on the same reel size.

A selection from a different range of inductors (Panasonic) is given in Table 6 for comparison, and these show a wider set of inductance values with differing characteristics. These inductors, with dimensions $4·5 \times 3·2 \times 3·2$ mm (length × width × height) are also obtainable in 8 mm tapes.

Table 6

Panasonic Part Number	Nominal Induct. (µH)	Tolerance (±%)	Q min.	DC Resistance max(Ω)	Min. Self-res. Freq. (MHz)	Allowable Current max(mA)
ELJ-FBR22M	0·22	20	30	0·20	230	700
ELJ-FB1R0M	1·0	20	30	0·44	80	470
ELJ-FB2R2M	2·2	20	50	0·59	45	410
ELJ-FB8R2M	8·2	20	50	1·6	24	250
ELJ-FB100K	10·0	10	50	1·8	22	235
ELJ-FB220K	22·0	10	50	2·6	15	195
ELJ-FB101K	100·0	10	40	8·8	6·7	105
ELJ-FB102K	1000·0	10	30	53·0	2·1	40

Note: Values are only a selection from the full range. See Published Data.

Tape packaging of these Panasonic SM inductors is at 8 mm-pitch positions in 8 mm tape; the 7-inch reel carries 500 pieces, and the 12-inch reel carries 2000 pieces. As with most SM products, bulk and magazine packaging are offered as alternatives, if required.

A simpler design of SM inductor consists of a wire coil wound directly on to a single-piece core former which is itself soldered directly on to the substrate. This is achieved by moulding the core former to a special shape in ferrite or in ceramic, the choice depending on the inductance required. The shape of the moulding is cylindrical in the centre so as to be able to carry the turns of wire, and has extended flattened 'feet' at the ends to provide two 'contact' areas. Each of these two contact areas is plated in such a way that the coil wire can be electrically joined to it, and the whole unit can be soldered on to the substrate.

Despite the simple construction of these inductors, the performance is comparable in virtually every way with that of the more complicated types. Similar inductance values with similar tolerances, Q-values, self-resonant frequencies, resistances and current-carrying capacities are provided by these units whose dimensions are generally more compact, that is, $2 \cdot 5 \times 2 \cdot 0 \times 2 \cdot 0$ mm including the winding.

Magnetically shielded miniature ferrite inductors for surface mounting are available from Siemens. A cylindrical drum-core with the winding is enclosed by a square-shaped base-plate and a square-shaped case-cap; the magnetically closed-circuit ferrite core reduces any stray inductive and magnetic coupling effects to a minimum. Almost cubic in shape, with dimensions of $4 \cdot 8 \times 4 \cdot 2 \times 3 \cdot 7$ mm (width × depth × height), the inductor has two soldering 'feet' each with dimensions $4 \cdot 5 \times 1 \cdot 7$ mm.

Inductance values range from $1 \cdot 0$ μH to 470 μH, each inductor having a Q-factor of 50. Across this range of inductance values, the self-resonant frequency falls from 150 MHz to 5 MHz, the resistance rises from $0 \cdot 16\Omega$ to 12Ω, and the current-carrying capacity falls from 790 mA to 45 mA.

3.11   ADJUSTABLE SM INDUCTORS AND TRANSFORMERS

Adjustable miniature inductors can be constructed from special cores which are obtainable for custom winding. Two ranges supplied by Siemens[14] permit the use of four terminals, thus making custom-designed transformers possible in these diminutive outlines.

A range of finished adjustable inductors in two sizes is available from Vanguard Electronics[15]. The 29000 Series has dimensions $0 \cdot 17 \times 0 \cdot 115 \times 0 \cdot 115$ in. and covers the range 18 nH to 120 nH with Q-factors falling from 52 to 22 over that range. Adjustment range is greater than ±10%, and the adjustment slot is situated on the side of the device. The 31000 Series has dimensions $0 \cdot 225 \times 0 \cdot 165 \times 0 \cdot 130$ in. and the inductance range is the same as that for the 29000 Series, but the adjustment range is ±25% and the adjuster is situated on the top of the device. Transformers, and fixed or variable tuned circuits built into small SM units are available from the same supplier.

3.12   CERAMIC FILTERS

Ceramic intermediate frequency filters for radio receivers are available in surface mounted form, and some examples are described in the following text.

The dimensions of a muRata ceramic IF filter for VHF AM/FM are

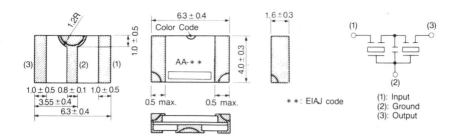

Fig. 3.18 Ceramic IF filter for selected intermediate frequencies around 1·07 MHz

$6·3 \times 4·0 \times 1·6$ mm. Each filter (Figure 3.18) has three contacts designed for surface mounting; there is a contact strip at each end (like those on the SM resistors) and another in the centre, all running the complete width of the device.

These filters are supplied with centre frequencies of 10·7 MHz (or there are several other centre frequencies near this value) with ±30 kHz tolerance, and a −3 dB pass-bandwidth of 280 kHz±50 kHz.

The muRata AM versions (which operate in the 'New Vibration Mode', and have the Registered Trade Name CERAFIL®) give a centre frequency of 455±2 kHz. With body dimensions $7·0 \times 4·5 \times 2·1$ mm, these filters have three small 'gull-wing' leads protruding from one side of the encapsulation for surface mounted soldering.

SM ceramic IF filters are currently deliverable in the standard 12 mm tape on reel, their longest dimension precluding the use of the standard 8 mm tape.

3.13   CRYSTALS AND OSCILLATORS

Miniature quartz crystals designed for surface mounting by vapour-phase reflow or wave soldering are made by Statek[16]. These devices have a low-profile hermetically sealed ceramic encapsulation with a soft-soldered glass lid (Figure 3.19(a)). Dimensions are $8·38 \times 3·94 \times 1·78$ mm (length × width × height), and the frequencies available range from 10 kHz to 2 MHz with calibration tolerances down to 5 ppm.

Miniature reference oscillators for use with clock-signal generators in microcomputer circuits, and in other control circuits, are also available in SM form. These are the two-terminal 'thickness-share mode' oscillators shown in Figure 3.19(b). There is a choice of centre frequencies from 3·0 to 6·14 MHz, with high Q and high stability.

The muRata product (Registered Trade Name, Ceralock®) is built in an encapsulation somewhat similar to the MELF design; there is a central portion with a square cross-section, and there are two end-contacts which are cylindrical with their diameter slightly greater than the length of the side of the central square cross-section. The components therefore show no tendency to roll when placed on a flat surface, and the cylindrical terminations permit good wetting by the solder during attachment to the substrate. Dimensions are $2·4 \times 2·4$ mm for the cross-section of the square central portion, 2·6 mm diameter for the cylindrical end contacts, and 6·9 mm over-all length. These dimensions permit packaging in standard 8 mm tape, and the devices can be placed at 4 mm-pitch positions in the tape.

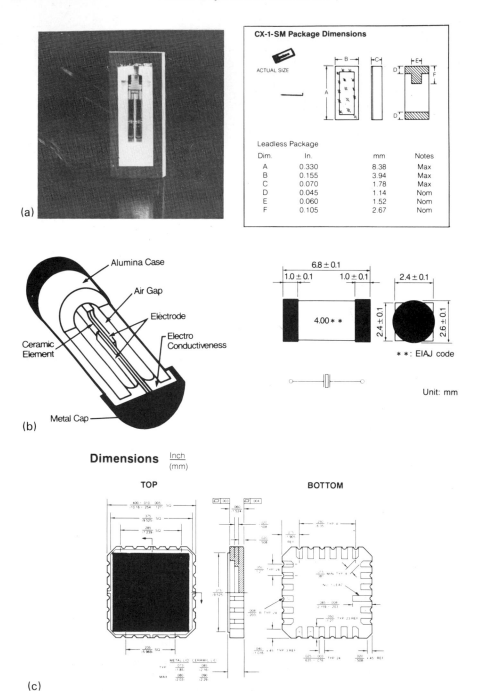

Fig. 3.19 (a) Miniature SM quartz-crystal (Courtesy of Statek); (b) high stability two-terminal oscillator with selected reference frequencies for microprocessor circuits around 3·0 to 6·14 MHz (Courtesy of muRata), and (c) 24-pin ceramic leadless chip carrier outline used for the Statek crystal oscillators. (Courtesy of Statek)

A range of crystal oscillators encapsulated in the standard 24-pin ceramic leadless chip carrier outline, shown in Figure 3.19(c), is also available from Statek. These are CMOS and TTL compatible and have frequencies up to 2 MHz.

3.14 SM POTENTIOMETERS

Adjustable potentiometers in SM form are available from a number of manufacturers. These are generally built (Figure 3.20) with all three surface-

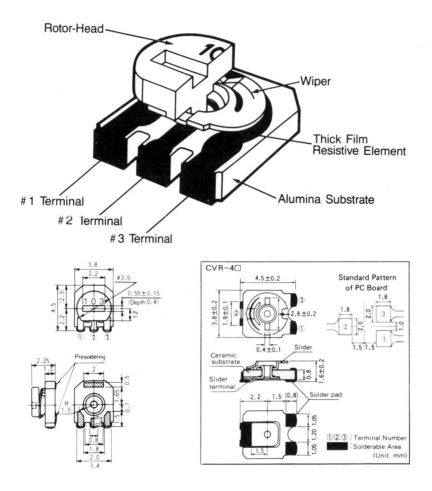

Fig. 3.20 Surface mounted adjustable potentiometers which withstand SMA soldering and cleaning techniques. (Courtesy of muRata at the top and bottom left and of Kyocera at the bottom right)

mounting contacts at one end of the body, but some types have the central contact at the opposite end of the body to the 'outer' contacts. Dimensions are typically $3 \cdot 8 \times 4 \cdot 5 \times 2 \cdot 3$ mm (width × length × height). About half the stand-off height of the whole unit is made up of the visible part of the adjustment cylinder which stands proud of the main part of the body.

Values extend from about 100Ω to 2·2MΩ with ±25% or ±30% tolerance on the full value of the resistance. The resistance temperature coefficients are of the order of ±250 ppm/K. Power rating is typically 0·1 W or 0·15 W, although some are rated at 0·3 W.

These potentiometers are capable of withstanding either immersion wave-soldering or reflow soldering; the close-fitting adjustment cylinder excludes molten solder entirely from the operational resistive film and the wiper during the soldering operation. Immersion cleaning in organic solvents also presents no difficulties.

Packaging of these diminutive potentiometers can be in tape and reel for those assembly machines which can handle these somewhat irregularly shaped units.

3.15 TRIMMER CAPACITORS

Trimmer capacitors are built into SM units with overall dimensions of about 4·5×4·0×3·0 mm (Figure 3.21). The lowest capacitance adjustment range of

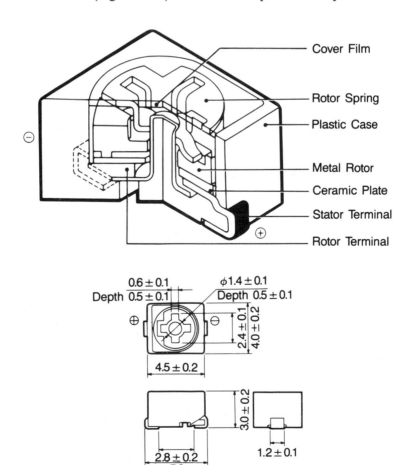

Fig. 3.21 Surface mounted trimmer capacitor which can withstand full SMA processing. (Courtesy of muRata, from whom alternative types of contacts can also be obtained)

the four available units is 2·0 pF to 6·0 pF, and the highest is 6·5 pF to 30 pF.

Trimming is effected by turning the central cylinder; this rotates a single asymmetrical plate which is isolated from the other (fixed) electrode by ceramic dielectric material in sheet form. The different minimum and maximum values of capacitance are obtainable in the same body-size with a consistent design using a single revolving plate; this is made possible by the manufacturer varying the type of ceramic dielectric to alter the capacitance value. The ceramics used are similar to those that form the basis of multilayer ceramic capacitors, being rated as NP0, N150, N750 or N1200. This means that the lowest values of trimming capacitance will have the lowest Temperature Coefficient (NPO, ±200 ppm) and the larger values of capacitance will have larger Temperature Coefficients.

Soldering can be by immersion or by reflow, and cleaning in organic solvents can also be applied within limits given in published data.

SM trimming capacitors which have been created by converting leaded devices are also offered by some suppliers.[17] These vary from single-turn horizontally-operated devices to multi-turn vertically-operated devices.

3.16 THERMISTORS

A number of circuits require resistive elements which have non-linear resistances with variation in temperature. These devices are generally referred to as Thermistors and are available in Surface Mounted form with both negative and positive co-efficients.

3.16.1 NTC

Negative Temperature Coefficient thermistors are available in SM form. The Thompson[18] range of 240 mW devices in the 1206 size range from 470Ω to 1 MΩ with B-values from 2800 to 4300. Resistance tolerance is ±10% or ±20% and the operating temperature range is −55 to +150°C. Soldering capability is up to 240°C for 5 seconds. A range of smaller devices in the 0805 outline is under development. A similar device in the 1005 outline is offered by Dale covering the resistance range 5 kΩ to 1 MΩ.

The Siemens range provides 100 mW devices ranging from a rated resistance of 2·2 kΩ to 100 kΩ, with tolerances between ±5% and ±20%, and with B-values in the region of 3530 to 4300 K. Cylindrical in shape, having dimensions 2·9 mm diameter and thicknesses between 0·7 mm and 3·0 mm, these NTC thermistors are attached to the substrate by soldering the two near-semicircular feet; these feet are separated by a small gap across the circular base of the device and have a narrow wetting area for the solder perpendicular to the base.

3.16.2 PTCs

Positive Temperature Coefficient thermistors in chip form are available from Siemens. Two ranges are currently offered.

The 0806 size devices, coded P381 to P401, have a maximum operating voltage of 25 volts and are intended for use as maximum temperature indicators. The rated threshold temperatures of the three types in the range are 110, 120 and 130°C.

The larger devices have a maximum operating voltage of 80 volts and are intended for use as protection elements. Three sized devices are offered: 1210,

1812 and 2220 with maximum operating currents of 100, 130 and 200 mA respectively. The devices are coded P2390—A707/8/9.

3.17 CONNECTORS

Virtually all electronic assemblies need connections to the outside environment for receiving input signals and power, and for driving external output functions. Connectors for SMD circuits have extra demands on them because SMDs allow assemblies to be extremely compact. Conventional leaded board connectors, if used in the most compact SMD circuit designs, mean that a disproportionate fraction of the board area is taken up by the connections.

Conventional leaded connectors can, however, generally be used with mix-print circuits. Mixprints have conventional components on one side of the board, together with SMDs on the other or both sides. Boards for mixprints can be either single-sided, double-sided or multilayer, but, because they include conventional leaded components, the boards will always be drilled for leads, and the spacing will be large enough to accept conventional components. Under these circumstances, a leaded connector can easily be included. The soldered leads give some rigidity, and additional screws or bolts complete the attachment. Such connectors are, however, unlikely to be suitable for automatic handling.

Exclusively surface mounted assemblies with SMDs alone will generally use either a multilayer board with fine tracks and buried vias, or some form of printed film circuitry on alumina or glass-coated metal. Holes will be undesirable and unnecessary for the main components, and spacing will be small. The inclusion of leaded connectors is therefore inconsistent with the general design of the substrate. If no holes at all are to be drilled, then connectors must be attached with either adhesive or solder or clamping, or by some combination of these methods.

The main problems of Surface Mounted Connectors are concerned with the stresses on the connections made between the connector pin and the substrate, the precision with which the connector has to be made and assembled and the materials chosen for its construction. These problem areas are tackled in different ways by the various manufacturers offering Surface Mounted Connectors and hence a range of answers have emerged.

One method of overcoming the problems is proposed by AMP[19]. The 'C' shaped compliant lead gives a compact design in respect of board area usage and effectively divorces the electrical conduction and mechanical connection by its length and 'C' shape. The mounting precision required is achieved by accurate manufacture and assembly of the metallic inserts and equipping the moulding with snap-fit pegs which will centralise within a locating datum hole. This feature provides some compliance along the length of the connector to allow for some 'float' during soldering, but gives more rigidity in the direction of mating, hence protecting the solder joints. The materials must be chosen with great care as the contact area of the hard solder joint is considerably smaller than in the case of a leaded connector. The housing material withstands the soldering temperature (215°C with vapour phase reflow) and the rapid increase from ambient to this soldering temperature normally seen in many reflow soldering systems.

ITT Cannon[20] use a 'J' tail similar to that used on PLCCs (see Chapter 4) to overcome the lead problems of SM connectors. The moulded body is similar to the standard D-type connector and rivets or bolts are required for fixing the device to the substrate via two holes[21]. Barbed retention posts forming part of the connector body will be used and variants which require no holes are being considered. An example of an ITT Surface Mounted Connector is shown in Figure 3.22.

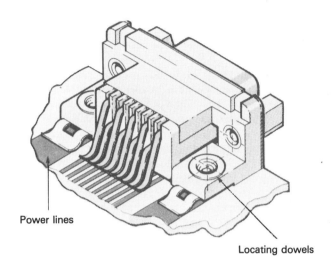

Power lines

Locating dowels

Fig. 3.22 Planned ITT Cannon surface mounted connector. (Courtesy of ITT-Cannon)

Fig. 3.23 Berg surface mounted connector. (Courtesy of Berg Electronics)

Berg[22] offers a line of Surface Mounted Horizontal Card connectors in both single- and double-row configurations, the contact floats in the housing allowing the legs to self-centre on the pads during soldering. The contact legs are basically of the 'gull-wing' shape as used by the SO Integrated Circuit package (see Chapter 4). Fixing holes are not incorporated in the moulded plastic bodies, mechanical strength being obtained by the double row of connection pads soldered to the substrate. Extra mechanical 'hold-downs' are available. To avoid damage to the legs of the devices, packing and transport to the pick-and-place machinery in tubes is advocated. Figure 3.23 shows both the single and double-row devices.

Dowty Electronic Interconnect[23] have an unusual approach to the design of high-density SM interconnection systems. With their detachable 'Snap' connector (Figure 3.24), connections are established between a special flexible circuit 'cable' (similar to a ribbon cable) with small raised contacts on the end, and the exposed copper tracks on the board. A retaining moulding is clamped to the board by a spring clip which passes through two holes drilled in the board and presses the end of the 'cable' on to the board. A sliding lock on the moulding tightens the spring in the engaged position after insertion, giving it the name 'Snap' which is the manufacturer's acronym for 'Sustained Necessary Applied Pressure'. Correct alignment between the flexible circuit 'cable' and the board is ensured by the two pillars, which are of different size and enter the two prepared holes in the board along with the clip ends of the retaining spring. The initial version of this connector has 40 contacts spaced at 40 contacts per inch. Others will have different numbers of contacts with the same spacing, or a higher density contact packing with 80 contacts per inch.

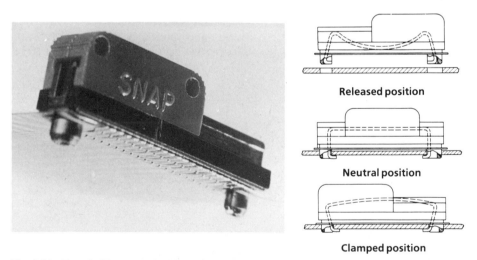

Released position

Neutral position

Clamped position

Fig. 3.24 Detachable Dowty 'Snap' surface mounted connector, and the method of action of its retaining spring and the sliding lock. (Courtesy of Dowty Electronic Interconnect)

A connector for surface mounting made by Hypertac[24] features 'door knob' terminations which are reputed not to spread the solder paste to the same extent as a 'blade'. The device needs no board drilling and may be used on double-sided printed circuit boards. It can be 'Picked and Placed' automatically and is compatible with vapour phase soldering techniques. The folded-under door knob version uses a smaller area of board and gives excellent alignment of the terminations with the tracks, at the cost of only 1·8 mm on the connector height (see Figure 3.25).

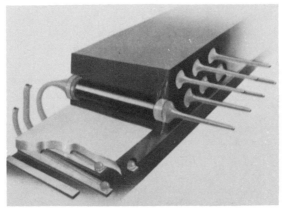

Fig. 3.25   Connector with soldered-under door knob terminations. (Courtesy of Hypertac)

3.17.1   Flex-rigid Prints

An alternative method of connecting totally surface mounted assemblies to external circuits is offered by the Philips Flex-rigid Prints, or F-R PRINTS.[25,26] Instead of constructing separate boards and connectors, the Flex-rigid approach is to make a continuous circuit board, part of which is rigid for carrying components while other parts are flexible for interconnection purposes. The 'junction' between the rigid component-mounted area and the flexible 'flying cable' area is therefore continuous for both the conductors and the substrate material, and there is no requirement for holes to carry separate connectors, or for any form of strengthening at the junction area. Additional rigid areas can be included at the remote ends of flexible flying leads for the attachment of virtually any form of external connector. External connectors can also be built directly on to the main rigid substrate area if required. A typical example of a customised Flex-rigid Print is shown in Figure 3.26.

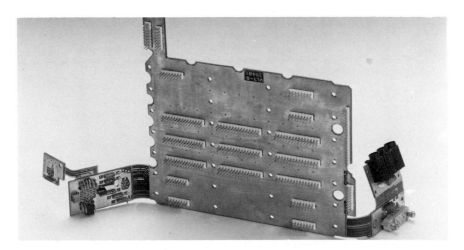

Fig. 3.26   Typical customised F-R PRINT or Flex-rigid Print. (Courtesy of Philips) [Ref. 24]

3.17.2　Elastomeric Connectors

A method of achieving high connection density for certain limited applications, and at the same time avoiding drilling holes for individual connections, is provided by elastomeric connectors. The simplest form of this type of connector is the inter-board connection[27]. The method of operation also applies to the variants, including, for example, one for mounting chip-carrier sockets.

The elastomer element is in the shape of a block which is retained by a bolted-on connector body (Figure 3.27), and across this element the electrical connections are made. Three types exist, as follows.

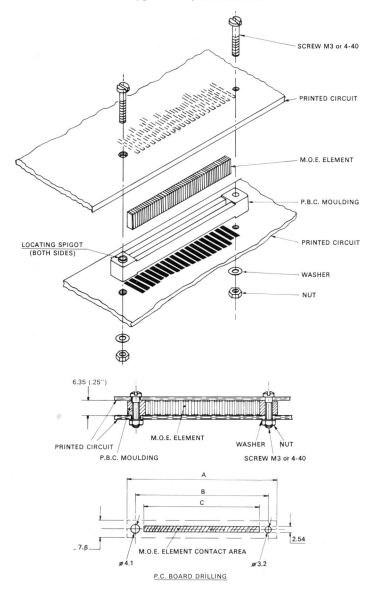

Fig. 3.27　Elastomeric connector for use between two printed circuit boards. (Courtesy of Flexicon Systems) [Ref. 25]

1 Laminated carbon, with alternate strips of insulating silicone and carbon-loaded silicone. This gives about 1Ω interface resistance per connection.
2 Laminated silver, designed as the laminated carbon version but with silver in place of the carbon. This gives about 0·1Ω per connection.
3 MOE, or Metal On Elastomer, with metal conductor strips on the elastomer core element. This gives about 0·02Ω per connection.

The contacts consist of the conductive laminations or strands passing from one side of the elastomer block to the other. The contact areas stand slightly proud of the remainder of the block, and the entire element is thick enough to ensure that the elastomer and the contacts extend from the connector body at both ends. When the connector is assembled, pressure applied by the bolts or other ties compresses the elastomer element. The contacts in the elastomer block are thus held in good contact with the copper pads or strips on the substrate. Simple connectors use bolts for this tensioning. The chip carrier uses a spring to pull against soldered ties, thus locating the four elastomer blocks (Figure 3.28) which

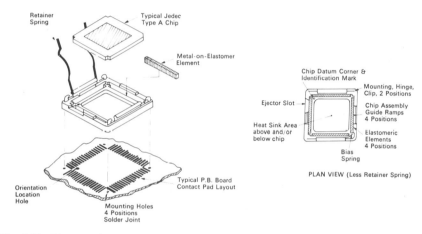

Fig. 3.28   Elastomeric connector for chip-carrier socket. (Courtesy of Flexicon Systems)

complete the connections between the board contacts and the contacts of the leadless chip-carrier.

3.17.3 Sockets

Surface mounted IC sockets can be obtained from Advanced Interconnections[28]. Each IC socket consists of a number of individual pin sockets, in a pattern suitable for the IC, attached to a Kapton sheet. Each pin socket has a flat base for attachment by reflow soldering directly to a substrate. The complete sheet of pin-sockets is attached and the Kapton sheet is removed after vapour-phase reflow soldering. The photograph in Figure 3.29 shows a typical board with sockets on both sides, both with and without through-connections.

3.17.4 Switches

DIP switches for surface mounting have been developed by CTS Corporation[29]. The packaging is designed to tolerate the reflow soldering process

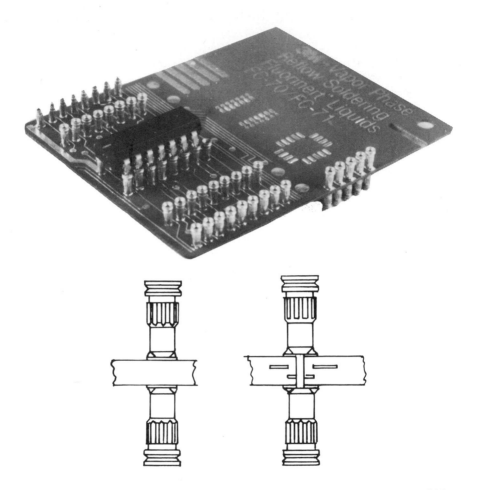

Fig. 3.29 Complete board with IC sockets made from individual single-pin sockets which are assembled in one step while mounted on a Kapton sheet which is later removed. Insert diagrams show socket pins connected through and not connected through. (Courtesy of Advanced Interconnections)

and the cleaning used with SMA. They are capable of being handled by automatic placement equipment and require no mounting pins for stability, as the low actuation force is not enough to break the solder bond. Types with both 'gull-wing' and 'J-lead' contacts will be made.

Augat manufacture the SMT/SMP series of subminiature toggle and pushbutton switches which are specifically designed for surface mounting.[30] They are available in single- or double-pole models in four function options. The switches feature all plastic, fully sealed bodies for good solderability and cleanability and offer a choice of gold- or silver-plated contacts. Both vertical or right-angle terminations are available together with a variety of toggle actuator lengths. They are suitable for automatic placement and can withstand the rigours of soldering at 250°C.

A range of 'DIL style' slide switches for Surface Mounting are provided by Dowty Electronic Interconnect. Resembling large SO integrated circuit packages (see Chapter 4), these devices offer Single-Pole-Single-Throw switches with from

2 to 12 poles in lengths of 4·95 to 30·35 mm. Body size is 7·2 mm wide by 3·1 mm high and the legs protrude to give an overall width of 9·8 mm. Recommended PCB pad sizes are 2·0×1·27 mm and spacing is 2·54 mm (0·1 in). The switches are rated at 100 mA at 30 volts DC and have a mechanical life of 50,000 on-off cycles.

Resistance to soldering heat is 10 seconds at 260°C and a top tape seal permits total immersion in most cleaning solvents.

3.17.5 Keyswitches and Keyboards

Individual keyswitches in surface mounted form are included in a complete keyboard from Cherry[31]. Although these keyswitches (Figure 3.30) are not

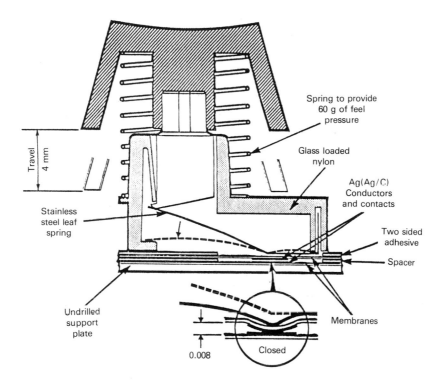

Fig. 3.30 Cross-section of a surface mounted keyswitch which is part of a keyboard. (Courtesy of Cherry).

available for general use, and the keyboard itself is not surface mounted, the use of these keyswitches indicates the stage of development of such products. Keyswitches and other small switch units will no doubt become available when there is sufficient demand and standardisation.

3.18 AVAILABILITY OF SM PASSIVE COMPONENTS

The major portion of this chapter gives an indication of the general availability of SM passive components; most kinds of passive components can be obtained either from one manufacturer or from several manufacturers. The components mentioned do not constitute an exhaustive list of what is currently available.

Further ranges of components are continually being introduced for surface mounted assembly. Some demand exists for virtually every type of component in surface mounted form. The general pattern is that one manufacturer will provide a close approximation to the device which is required, and the various market influences will then cause 'second-source' suppliers to appear. This will be followed by other comparable products and by increasing standardisation and rationalisation.

The intending user of Surface Mounted Assembly can rest assured that by examining the literature from the component manufacturers, the required component will almost certainly be found. In many cases the manufacturers encourage discussions about the possibility of 'customised' selections of devices, or even customised devices, where this is of mutual commercial benefit.

The entire concept of SMA draws the SMD manufacturer and the equipment manufacturer into closer and more profitable consultation with one another, whether it be to define packaging requirements or to devise a new component. Against this background, the equipment manufacturer can initiate designs with the confidence that the appropriate SMD can be obtained.

REFERENCES

** 1 'Surface Mounted Devices', various publications Philips Electronic Components and Materials Division Building BA, Eindhoven, Netherlands. Tel: 040 723331.

** 2 Matsushita Electric Trading Co. Ltd, Electronic Components Dept, PO Box 288, Central, Osaka, Japan. Tel: 06 282 5111.

** 3 'Surface Mounting Components Catalog', Dale Electronics Inc., 2300, Riverside Blvd, Norfolk, NE 68701, USA. Tel: 402 371 0800.

4 'Multi-Layer Ceramic Capacitors', Data Sheet, Kyocera Corporation, 52-11, Inoue-cho, Higashino, Yamashina-ku, Kyoto 607, Japan. Tel: 075 592 3851.

5 'Multilayer Ceramic Capacitors', Technical Bulletins Nos. 14 and 15, 1983, AVX Corporation, Myrtle Beach, SC 29577, USA. Tel: 803 448 9411.

6 'High Voltage Ceramic Chip Capacitors', Data Sheet, Novacap, 1811, North Keystone Street, Burbank, CA 91504, USA. Tel: 213 841 7150.

** 7 'High Voltage Capacitors', Brochure, Semtech Corporation, 652, Mitchell Road, Newbury Park, CA 91320, USA. Tel: 213 628 5392.

** 8 Sprague Electric Company, North Adams, Massachusetts 01247, USA. Tel: 413 644 4411.

** 9 'Surface Mounted Components—Engineer's Guide', Wimpey-Dubilier Limited, Launton Road, Bicester, Oxon, OX6 0TU, UK. Tel: 0869 242035.

10 'Metallised Polyester Chip Capacitors—Provisional Data', Waycom Limited, Wokingham Road, Bracknell, Berkshire, RG12 1ND, UK. Tel: 0344 24571.

**11 'Leadless Aluminium Electrolytic Capacitors', Data Sheet, Nippon Chemi-Con Corporation 7-8, Chome, Yukato-cho, Shinagawa-ku, Tokyo, Japan.

**12 Murata Manufacturing Co. Ltd, 26-10 Tenjin 2-chrome, Nagaokakyo, Kyoto 617, Japan. Tel: 075 921 9111.

13 'Surface Mount Inductors Series 1008', Data Sheet, Coilcraft, 1102 Silver Lake Road, Cary, Illinois 60013, USA. Tel: 312 639 2361.

14 'Components for Surface Mounting', Short Form Catalog 1984/85, Siemens AG, Bereich Bauelemente, Balanstrasse 73, Postfach 801709, D-8000 München 80, FRG. Tel: 089 4144-1.

15 'Hybrid LC Filters', Brochure, Vanguard Electronics Co. Inc., 1480 West 178th Street, Gardena, CA 90248, USA. Tel: 213 323 4100.

16 'Surface Mountable Crystal Oscillators & Miniature Quartz Crystals', Preliminary Data Sheets, Statek Corporation, 512 N. Main, Orange, CA 92668, USA. Tel: 714 639 7810.

**17 Bourns Inc., Trimpot Division, 1200 Columbia Avenue, Riverside, CA 92507, USA. Tel: 714 781 5050.

18 'Thermistors for Industrial Applications', Data Sheet, Thomson-CSF Composants, 101, Bd Murst 75781, Paris, Cedex 16, France. Tel: 331 743 9640.

19 'Factors Affecting Surface Mount Technology', *New Electronics*, 25th June (1985).

**20 'ITT Cannon Surface Mount the Connector', *Electronic Engineering*, August (1984). Cannon Electric (GB) Limited, Days Close, Viables Industrial Estate, Basingstoke, Hants, RG 22 4BW UK.

21 'Interconnection—Meeting Needs to Keep Pace with Technology', *Electronics Weekly*, 26th June (1985).

22 'Surface Mount—the Berg Connection', *Electronic Engineering*, March (1985). Du Pont de Nemours International SA. 50-22, route des Acacias, 1211 Geneva 24, Switzerland. Tel: 022 378111.

**23 'Two Interconnection Innovations', *Electronic Product Design*. November (1984). Dowty Electronic Interconnect, High Wycombe, Tel: 0494 26233.

24 'Approaches to the Surface Mounting of Connectors', Hypertac Limited, Chronos Works, North Circular Road, London, NW2 7JT, UK. Tel: 01 450 8033.

25 'Flex-rigid Prints' Mullard Limited, Mullard House, Torrington Place, London, WC1E, 7HD, UK. Tel: 01 580 6633.

26 'Flex-rigid Prints—the Cost-cutting Interconnection System', Mullard Technical publication M83-0193, Mullard Limited, Mullard House, Torrington Place, London WC1E 7HD, UK. Tel: 01 580 6633.

**27 Spencer, A., 'Elastomeric Connectors— A Flexible Friend', Flexicon Systems' *New Electronics*, 26th June (1984). Flexicon Systems Limited, Hitchen Street, Biggleswade, Beds, SG18 8BH, UK. Tel: 0767 312086.

28 'Surface Mounted Sockets—Low Insertion Force, Preliminary Data Sheet, Advanced Interconnections, 5, Division Street, Warwick, RI 02818, USA. Tel: 401 885 0485.

29 'Surface-mounted DIP Switches Withstand Tough Treatment', *Electronic Design*, p. 30 March 7, (1985). CTS Corporation, 905 N. West Boulevard, Elkhart, IN 46514, USA. Tel: 219 293 7511.

30 'Switching to the Surface', *Electronic Equipment News,* November (1985).

31 'Surface Mounting Moves to the Keyboard', *Electronic Engineering*, December (1984), Cherry Electrical Products, Coldharbour Lane, Harpenden, Herts, AL5 4UN, UK. Tel: 05827 63100.

Chapter 4

ACTIVE SURFACE MOUNTED DEVICES

E. G. EVANS and G. A WILLARD*

EDS, Cheam, Surrey, England
*Mullard Mitcham, Mitcham, Surrey, England

4.1 INTRODUCTION

During the 1950s, the introduction of the two hybrid assembly technologies, Thick Film and Thin Film, created the need for surface mounted semiconductors. Both of these technologies were able to produce conductor tracks, resistors, small-value inductors, and a limited range of capacitors. In both technologies, however, it was necessary to add discrete active devices such as transistors and diodes after the majority of the passive components had been fabricated on the substrate. As part of the process of adding these discrete encapsulated semiconductors, holes were required in the substrate to accept the leads. The formation of holes in the alumina or other hard material was difficult and almost prohibitively expensive. Although this assembly procedure was used for a period, the difficulty and the cost naturally motivated designers to find satisfactory ways of attaching the discrete semiconductor to the substrate without the need for the holes. The following three main ways were used.

The first method, which offered the greatest variety of readily available semiconductor devices, was simply to take the standard fully-encapsulated semiconductors which were available for conventional printed circuit assemblies and to solder these directly onto the contact areas of the substrate without holes. Soldering by any method was difficult because the lead-pins of the devices were not designed for attachment in the absence of holes. Furthermore, the miniaturisation of the end product was jeopardised by the size of the components. Only the fact that this was the least expensive method as well as the one which gave unlimited access to the full range of available devices enabled it to survive for any length of time.

A second possibility, which preserved the miniaturisation, was to use unencapsulated dice or chips. These 'naked chips' were attached by bonding the back of the chip to the substrate. Thin wire leads were then attached to the small contact pads at the front surface of the chip and to the adjacent contact pads on the surrounding area of the substrate. The device was protected by a lacquer or epoxy coating which was added after the bonding of chip and connections was completed. This method of surface mounting the device required a level of skill in assembly comparable to that of the manufacturers who made encapsulated devices. It reduced the number of devices which the equipment assembler could immediately and easily obtain. The level of skill and technology which was required meant that this approach was not an economically attractive proposition for many equipment assemblers. The technique is still current, however, among specialist equipment manufacturers making such products as analogue-to-digital or digital-to-analogue converters and various types of modules for military applications. It is particularly useful where power semiconductors are required. More is said of this technique later. Several alternative compromise solutions were pursued, attempting to combine the miniaturisation of 'naked chips' with a more practical assembly method so that there would be a larger number of users. These approaches included semi-naked chips, such as the solder-bump flip-chip, and more recently, the Tape Automated Bonding (TAB) assembly of chips on a polyimide tape, specially fabricated by the device manufacturer for a closely defined market consisting of the very limited number of users.

The third possibility was for the device manufacturers to encapsulate the chip in a much smaller outline with contacts or leads specifically designed for surface mounting by direct and easy soldering to contact pads on the substrate, thus making a true 'surface mounted device'. The first commercially released surface mounted transistor encapsulation was the SOT-23. This outline appeared in the late 1960s and has been marketed successfully ever since then, along with many other now standard outlines for both discrete semiconductors and integrated circuits.

In the encapsulation identification number SOT-23, the letters stand for Small Outline Transistor. The letters SO are also used as part of one series of integrated circuit outline identifications. The result is that the letters SO have become almost a generic term used colloquially for all such devices. SO devices are now accompanied by other device encapsulations based on similar concepts, and these have led to the full range of currently available surface mounted discrete semiconductors and integrated circuits which are described in more detail later in this chapter.

4.1.1 Unencapsulated Chips

The technology of attaching unencapsulated dice or chips to a substrate remains an important means of creating miniaturised electronic circuits, although it can not be classified as an ordinary 'surface mounting' technology. Compared with the mainstream surface mounting methods, the assembly procedures for naked chips require a bigger investment in gaining expertise, but when the procedure is properly applied the result can be even more compact circuitry.

Virtually all of the chips for discrete semiconductors can be obtained in this 'naked chip' form. However, the larger power chips and high voltage chips are the most widely used in this form because, generally, these are the only alternatives to large full-size encapsulated devices.

An insight into the naked chip technology helps to underscore the attractive simplicity of mainstream surface mounting procedures. When semiconductor and integrated circuit chips are made, they are formed on the surface of a silicon slice generally having a diameter of several inches (Figure 4.1). The chip's circuit is created by a series of processes which are step-repeated at precise intervals so that the top surface of the slice becomes populated with a large number of completed chip circuits. These chip circuits are tested by probing each 'chip' individually. Unsatisfactory chips are marked with ink. The chips are later separated by

Fig. 4.1 Semiconductor slice prior to division into dice. (Courtesy of Philips)

scribing or laser cutting. Acceptable chips are mounted on a lead frame and then encapsulated.

As an alternative to these later stages, the slice or the separated chips can be supplied to an equipment assembler who can then mount the 'naked chips' on his own substrate. The equipment assembler must also then be able to build an accurately placed set of miniature jumper leads which connect the minute contact pads on the chip to the small contact areas on the substrate. Power chips or any high voltage chips which have been specially constructed for this method of assembly will generally have been glass-passivated over the non-contact areas by the manufacturer so as to provide partial protection. A form of encapsulation is then applied after the chip has been installed on the substrate. This is generally a layer of epoxy or lacquer. Alternatively it can be a specially designed cap or lid which is attached to the substrate, covering the chip.

Semiconductor manufacturers such as Sprague[1] specialise in offering a very wide range of chips for this kind of application, and all of the larger manufacturers offer both standard and customised chips with a variety of metallisations for bonding and connecting. Gold backing for eutectic bonding is frequently offered for highly reliable attachment. Die temperature during attachment is permitted to rise to 425°C for a period of up to two minutes. Thus the intending user of naked chips has little difficulty in obtaining the chips required, and the first stages of attachment are relatively easily accomplished. Wire bonding to the contacts for accessing the chip, however, requires considerable specialised skill and costly machinery.

Advantages of using naked chips can include reduction in the size of the final product, slightly enhanced high frequency performance because of the reduced lead-length, and reduced cost on lengthy production runs if implementation is successful. Disadvantages include the requirement for more fabrication expertise and equipment. There is also the risk of increased costs if assembly production runs are not successful.

4.1.2 Mixing Technologies in Mixed Prints

Because, as mentioned earlier, several assembly technologies exist concurrently, it is feasible to mix several different styles of device in the same 'mixed print' or 'mix-print' execution. The surface mounted active devices can, in principle, be employed alongside other encapsulated devices which are inserted into holes, or they can even be mixed with naked chips assembled directly on to the substrate.

Currently, the mixed print execution is valid for the simultaneous assembly of dedicated surface mounted devices, both active and passive, alongside those specialised passive devices for which holes are still required.

The descriptions which follow apply exclusively to what can be regarded as standard discrete surface mounted devices. These are encapsulated and have solderable contacts or leads. Mixed-print technologies and the use of naked chips are not discussed further in this Chapter.

4.2 ACTIVE SURFACE MOUNTED DEVICES—GENERAL FEATURES

A wide range of active SMDs is now available. All active SMDs are small compared with their conventional equivalents. They can withstand the high temperatures associated with soldering. They can be soldered on either side of the board with their purpose-designed soldering contacts. These devices are sometimes collectively called 'leadless', but this is erroneous because many have short flat leads specially shaped for surface mounting. The short surface mounting leads on some of the devices give a certain amount of mechanical flexibility. This is particularly helpful with the larger SMDs where a mismatch in thermal coefficient of expansion between component and substrate might otherwise cause stress which could possibly reduce component life in certain circumstances.

The use of active SMDs is fully consistent with the use of passive SMDs. Many active SMDs are supplied in the same kind of tape and reel packaging as used with passive SMDs. The same 'design rules' are necessary to ensure that substrates are compatible with devices. Substrates must be prepared for receiving the device, either with adhesive or solder paste, depending on the assembly method (see Chapter 7). Automatic assembly machines can be used for removing components from packaging, where this is applicable, and for mounting the devices on the prepared substrates (see Chapter 6). Appropriate soldering methods have to be employed. Suitable inspection methods and quality standards are required both for the devices and for the assembled boards. Dimensions of SMDs are typified by the surface mounted components shown in Figure 4.2 (a) discrete semi-conductors and (b) integrated circuits.

(a)

(b)

Fig. 4.2 Typical semiconductor SMDs. (a) discretes; (b) integrated circuits. (Courtesy of Mullard)

4.3 DISCRETE SM SEMICONDUCTORS

Because crystals which are used in small encapsulations for surface mounted devices are generally identical to those which are used for the conventional larger encapsulations, the electrical performance of comparable surface mounted and conventional devices will always be virtually the same. If there is any difference, this will normally be attributable to the shorter lead length in the surface mounted version, the slightly lower inductance giving a marginally improved high frequency performance. The power dissipation capabilities of surface mounted devices will, however, be more restricted than those of conventional devices because of the reduced physical size.

Since the chips are encapsulated, automatic testers can effect full electrical testing by applying external probes to the leads or contacts in the same way as with the larger conventional component. The probes can be applied to the fully powered-up device to test all working parameters before assembly on to the substrate, and the relevant working parameters after assembly. This contrasts with the 'naked chip' which cannot be fully tested electrically by probe. Thus, although the surface mounted device is only marginally larger than an unencapsulated chip which has been 'covered' or protected by the epoxy or lacquer layer, it offers all the advantages of the robustness and testability of a conventional device.

4.3.1 The SOT-23 Encapsulation

The first commercially available miniature encapsulation was the plastic SOT-23, which was introduced in the late 1960s. It was designed primarily for transistors and therefore had three solderable contacts. Diodes were also made available in the same encapsulation, one or more diodes using either two or three of the contacts. A wide range of low-power transistor chips and diode chips soon became available in the SOT-23 outline.

The appearance of the SOT-23 encapsulation is shown in Figure 4.3 and the

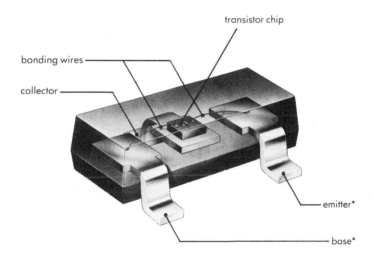

Fig. 4.3 See-through view of SOT-23 encapsulation. (Courtesy of Philips)

dimensions are given in Figure 4.4. The Free Air power dissipation is normally limited to 200mW for a T_{jmax} of 150°C with normal unrestricted convection in an ambient temperature of 25°C. When, however, the SOT-23 is attached to an alumina substrate, the total power dissipation can be raised to 350mW at an ambient of 25°C, derating by 2·8mW/K to 150°C which is the maximum operating or storage temperature. The thermal resistance from junction to ambient is usually given as 625K/W.

The 'gull-wing' leads of the SOT-23 visible in Figure 4.3 permit easy access by test probes for checking the component both before and after it has been soldered to the substrate. The small penalty incurred by this design, compared with

'J-bend' connections which 'tuck under' the body of some of the integrated circuit encapsulations, is the slightly enlarged 'footprint' caused by the extended contacts.

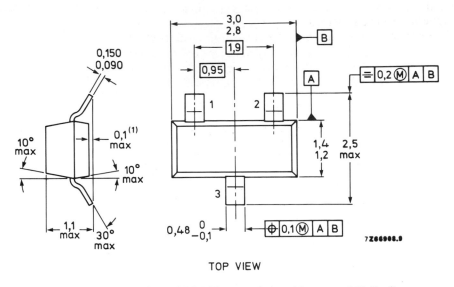

TOP VIEW

Fig. 4.4 Dimensions of SOT-23 encapsulation. (Courtesy of Mullard)

The SOT-23 continues to be one of the most 'popular' encapsulations for transistors and diodes. A very extensive range of device types is available from virtually all of the semiconductor manufacturers. Some have been sold under trade names such as 'MiniBloc'[2] from Motorola, and 'Micromin'[3] from Philips, the latter covering virtually every type of standard low-power and medium-power transistor and diode chip. Many other suppliers also list a wide selection of devices. Ferranti[4], for example, list general purpose transistors, medium power transistors, switching transistors, v.h.f. transistors, high voltage transistors, darlingtons, silicon planar high-speed switching diodes, silicon ion implanted hyperabrupt tuner diodes, Schottky barrier diodes, and zener diodes.

In addition to the standard pinning versions, some manufacturers also make reverse pinning versions available at special request. The multi-sourcing of such a large variety of frequently used types suggests that this encapsulation will continue to be used almost as an industry standard for the foreseeable future.

Automatic placement machines, both small and large, are also readily available for devices in the SOT-23 encapsulation and the other common encapsulations which are described below. This point should be borne in mind when considering the more recent and more elaborate encapsulations. The less 'standard' encapsulations are supported by fewer suitable placement machines.

4.3.2 The SOT-89 Encapsulation

In the mid-1970s, the SOT-89 followed the plastic SOT-23. It had a slightly larger but fully re-designed three-terminal outline. This was capable of dissipating more heat, particularly through the exposed metal central contact.

The contacts of the SOT-89 encapsulation are not 'gull-wing' like those of the SOT-23. They are thicker and emerge directly from the encapsulation level with the bottom of the plastic. Being thicker and straight, they permit improved heat flow from the chip to the outside. The central contact has a wide, more or less rectangular, exposed area beneath the centre of the encapsulation. This contributes very considerably to the good heat-removal properties by direct conduction to the substrate.

The appearance of the SOT-89 encapsulation is shown in Figure 4.5 and the

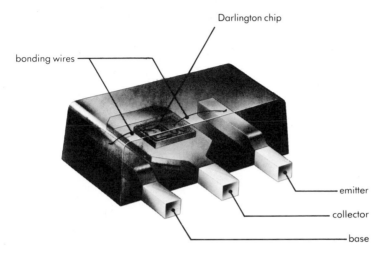

Fig. 4.5 See-through view of SOT-89 encapsulation. (Courtesy of Philips)

dimensions are given in Figure 4.6. The Free Air power dissipation is generally given as 500mW for a T_{jmax} of 150°C with unrestricted convection in an ambient temperature of 25°C, this being clearly greater than the earlier SOT-23. When the SOT-89 is attached to an alumina substrate, the total power dissipation can be raised to 1·0W at an ambient of 25°C, derating by 8·0mW/K to 150°C, which is the maximum operating and storage temperature. Thermal resistance from junction to ambient is usually given as 250K/W, a considerable improvement on the SOT-23.

The SOT-89 is the most comonly used outline for the higher power devices, and like the SOT-23 it continues to be virtually an industry standard encapsulation. A wide range of chips are available in the SOT-89 encapsulation from various manufacturers. Most of the above-mentioned manufacturers who provide SOT-23 devices also offer SOT-89 devices. For example, Philips[5] and their associated companies such as Mullard[6] in the UK and Amperex in North America, provide an extremely large variety of chips in this encapsulation. As well as the standard pinning types, reverse pinning types can also be obtained, if required, from some sources. Automatic placement machines for handling SOT-89 are also widely available.

4.3.3 The SOT-143 Encapsulation

In the early 1980s, the plastic SOT-143, a surface mounted four terminal outline, also appeared. The SOT-143 encapsulation is shown in Figure 4.7 and the dimensions are given in Figure 4.8. Free Air power dissipation is normally limited to 200mW, identical to the SOT-23. The maximum dissipation is generally the

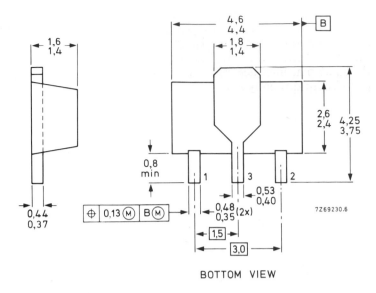

Fig. 4.6 Dimensions of SOT-89 encapsulation. (Courtesy of Mullard)

same as the SOT-23 under all comparable conditions.

The SOT-143 is the most commonly used outline for devices needing the four connections, such as p-n-p-n trigger devices and some FETs. Like the SOT-23 and SOT-89, it continues to be virtually an industry standard encapsulation. The number of chips available in SOT-143 from various manufacturers covers the full

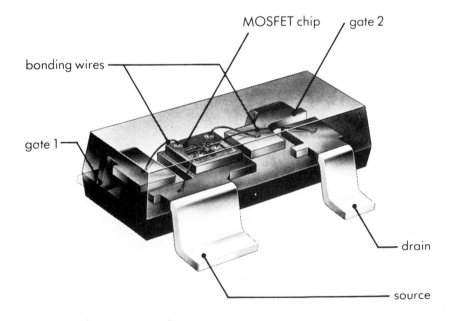

Fig. 4.7 See-through view of SOT-143 encapsulation. (Courtesy of Philips)

range of device chips which are available in conventional encapsulations, although that number is, of course, small compared with SOT-23 and SOT-89 devices. Reverse pinning types can also be obtained from some sources, if required. Automatic placement machines for handling SOT-143 are also widely available, the same machines generally being able to handle both SOT-143 and SOT-23 types without difficulty.

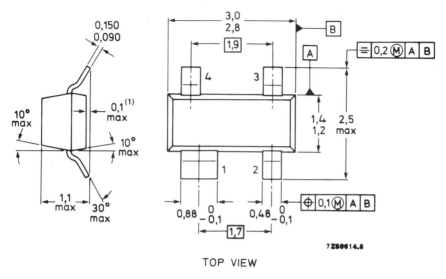

TOP VIEW

Fig. 4.8 Dimensions of SOT-143 encapsulation. (Courtesy of Mullard)

4.3.4 Cylindrical Diodes

When any single diode is built into the SOT-23 encapsulation, one of the three contacts is strictly redundant. This is true whether or not the third contact is used as an alternative connection to one pole of the diode. Since large numbers of single diodes are used in circuit assembly, and each diode properly requires only two contacts, there is a well-defined need for a simple two-contact diode encapsulation. Two such diode outlines, each cylindrical with contacts at the ends of the cylinder, are described below.

4.3.4.1 THE SOD-80 DIODE ENCAPSULATION

The first two-contact diode outline appeared a short time after the advent of the SOT-143. This was the SOD-80 encapsulation, of similar proportions to the MELF resistors but appreciably smaller. Diodes produced in this outline are called 'MiniMELF' by one manufacturer (ITT). Specifically designed for small diode chips, the SOD-80 encapsulation is less costly than SOT-23. It uses less board space, and it also weighs less than the SOT-23.

The Philips version of the SOD-80, a hermetically sealed glass-and-metal structure, is shown in Figure 4.9 and the dimensions are given in Figure 4.10. Many types of diodes both for small signal functions and for low-power rectification are available in this outline. A range of nitride-passivated zener diodes is also obtainable from Motorola. Operational power dissipation for this

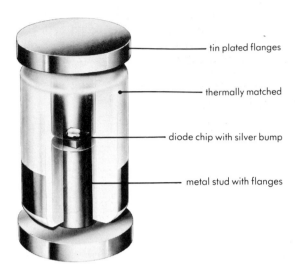

Fig. 4.9 See-through view of SOD-80 encapsulation. (Courtesy of Philips)

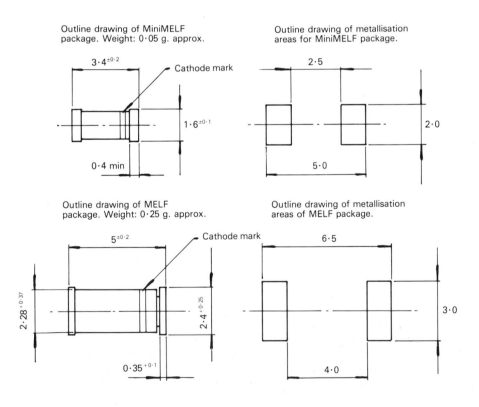

Fig. 4.10 Dimensions of SOD-80 encapsulation (called 'MiniMELF' by ITT) and the 'MELF' diode encapsulation. (Courtesy of ITT)

encapsulation is generally limited to 250mW, or in some instances, 500mW. The length of 3·5 mm and the diameter of 1·6 mm limit the power handling capability and so, when the device manufacturer needs to circumvent this limitation, the larger cylindrical encapsulation which is described in the next section is used.

Packaging of SOD-80 diodes for automatic placement in a surface mounted assembly is normally in 8 mm tape and reel.

4.3.4.2 THE MELF DIODE ENCAPSULATION

A larger cylindrical diode encapsulation is used for chips which need to handle either considerably higher voltages or more power than can be permitted with the SOD-80 encapsulation. With length 5·0 mm and diameter 2·4 mm as shown in Figure 4.10, the dimensions are similar to those of the MELF resistors. For this reason, the encapsulation is called the 'MELF' diode by one manufacturer (ITT)[7]. With adequate cooling, power dissipation can be as high as 2·0W for some specific glass-passivated power rectifier chips.

For rectification, maximum recurrent peak reverse voltages and DC blocking voltages up to 1000 V are obtainable in, for example, the General Instruments Superectifier[8] range for which the execution of this encapsulation is in moulded plastic on glass between the metal terminations. These offer maximum average forward currents of 1·0 A, and peak forward surge currents of 30 A.

Packaging of these larger diodes for automatic placement in surface mounted assembly is in 12 mm tape and reel.

4.3.5 Conventional Outlines Adapted for SMDs

In recent years, the market for surface mounted devices has become fully established. This enlarged market for surface mounted devices has stimulated demand for additional types. It has also correspondingly encouraged manufacturers to believe that tooling up for such new surface mounted products is a worthwhile investment. In particular, some 'new' surface mounted outlines have been produced by adapting a number of conventional semiconductor outlines.

It is possible to modify an existing encapsulation if it has contacts emerging in such a way that they can be formed into a single plane of solderable contacts around the device. This is true for several existing outlines and, in appropriate commercial circumstances, these could become available in modified form. The major advantage of following this procedure is that devices with considerably higher power capability than standard surface mounted devices can be made available.

In particular, several manufacturers have supported this trend by offering a range of TO-126 and TO-220 devices in a modified form for surface mounting. In both cases the flat leads which emerge from one side of the device have been re-shaped and shortened to bring all of them to suitable soldering contacts in the same plane as the heatsink of the encapsulation. The wire-leaded E-line devices have also been made available in surface mounted form (known as the 'E-line joggle')[9] by Ferranti by bending the shortened leads into the same plane as one of the flat external surfaces.

One factor which has to be considered is that the devices produced in this way are not always capable of being handled by currently available surface mounting placement machines.

4.3.6   Performance Obtainable with Semiconductor SMDs

An extremely wide range of surface mounted device types, both diode and transistor, is now available. The performance available to the circuit designer is limited only, therefore, by the ability of the device manufacturer to build the required chip into an acceptable encapsulation. Except for the higher power devices, this presents no problem, most low-power chips being available from at least one of the device manufacturers.

Devices include, for example, high frequency diodes and transistors with f_T as high as 5 GHz, and high voltage devices with V_{CEO} up to 300 V. Low-noise devices are available, as are fast-switching, variable capacitance, and voltage regulation devices. The range of semiconductor technologies encompasses bipolar, MOS, junction FET, and Schottky.

Thermal characteristics governing the maximum power which can be dissipated are affected by the size and the encapsulation. The power rating which is likely to be stated for any given chip is therefore limited by the encapsulation chosen by the manufacturer. The method of mounting the encapsulated device on the substrate, the ambient temperature, and the method of cooling the device will all affect the actual power dissipation. The Data Sheets should be consulted for all factors that relate to the use of each of these devices.

4.3.7   Soldering Semiconductor SMDs

Soldering of all these semiconductors by standard SM immersion or reflow methods is possible. For immersion soldering, they can generally be subjected to 260°C for 10 seconds, although some manufacturers limit this to 5 seconds. Even this shorter period is still adequate for the dual-wave immersion soldering under normal conditions. Some specific device types are capable of withstanding 280°C for 10 seconds, but the published Data Sheets should always be consulted. Clearly, the actual soldering conditions must simultaneously satisfy all of the requirements for all of the devices on the board if the soldering operation is to be successful. In the subsequent servicing or repair of operational boards, the application of such methods as hot-iron soldering or localised hot-air is feasible provided that the temperature-time limits which are stipulated in the Data Sheets are strictly observed. For terminations alone, temperature limits as high as 450°C for 10 seconds are stated in some instances as acceptable during localised hot-iron re-soldering. Clearly, no assumptions should be made about these elevated temperature limits, and if in doubt the manufacturer should be consulted.

Most soldering contacts have a finish coating of Sn-Pb solder. As well as facilitating storage for a considerable period in non-hostile atmospheres, the finish also ensures good soldering if conventional hot Sn-Pb soldering is used.

Occasionally, the use of conductive adhesive is employed to attach active devices to a board which has already been built up with passive components. A silver-filled epoxy permits the addition of complex encapsulations without raising the temperature to the same levels as are required for soldering. Details of suitable adhesives will generally be available from the device manufacturer. With these details there will also be given information concerning temperatures to be applied and the hardening times which are necessary.

After assembly of the boards, cleansing agents such as isopropyl alcohol or Freon are normally classified as satisfactory for these devices. The published

Data Sheets should nevertheless always be checked for confirmation of suitability of any proposed cleansing agent.

4.3.8 Packaging Semiconductor SMDs

Tape and reel packaging is used for delivery to automatic placement machines (Figure 4.11). Standard 'Super-8' tape is employed where possible for SOT-23

(a)

(b)

Fig. 4.11 Tape and reel for transistors and diodes. (a) metal tape with SOD-80 diodes; (b) close-up of plastic tape with SOT-23 transistors. (Courtesy of Philips)

and SOT-143, but SOT-89 devices are generally delivered in 12 mm tapes. The 7-inch reel carries up to 4000 plastic-encapsulated semiconductors or SOD-80 diodes. Reels with 10,000 devices can be ordered from a few manufacturers. All MOS devices are delivered in an antistatic version of the tape. Some manufacturers permit the customer to choose the orientation of the devices that he requires within the tape. For example, Motorola[2] offer what is called orientation 'T1' for SOT-23 with the single contact on the left of the tape determined from the direction of the tape feed. Orientation 'T2' has the small contact of SOT-23 reversed, that is on the right as the tape travels forward.

Bulk packaging is frequently in plastic tubes with detachable lids for quantities of up to 1000. Plastic bags and boxes are also used for delivery to hopper feed machines.

4.3.9   Reliability of Discrete Semiconductor SMDs

Surface mounted transistors and diodes have gained, over many years, a well-earned reputation for reliability. Manufacture of the majority of the devices in the wide range which is now available can be said to be based on well established techniques. Long-term marketing of SMDs on a world-wide basis by the major component manufacturers means that today's products can be said to possess an extremely respectable pedigree.

Most of the early applications of surface mounted devices were either in professional or military equipment. For this highly specialised market gold-to-gold chip bonding was invariably used. As commercial interest in the new small encapsulations widened, cost reductions became desirable. Justifiable economies were achieved by changing to the now well-established gold-to-aluminium bonding. Because production techniques had also improved as a result of constant research and development on the part of the larger manufacturers, only a very modest reduction in long-term reliability was incurred by the change to gold-aluminium bonding, and this small loss could be regarded as negligible by the majority of users in the new market. Manufacturers who use the Pro Electron[10] type-numbering system sometimes indicate their use of gold-to-gold bonding with three letters such as BCW followed by two digits. Gold-to-aluminium bonding is then frequently indicated by two letters such as BC followed by three digits. A typical example of the use of this method of indicating bonding is given by the surface mounted npn transistors BCW32 in gold-gold bonding and BC848 in gold-aluminium bonding. Both are near-equivalents of the conventionally mounted BC548 or BC548B in TO-92 outline, or the BC108 in TO-18 outline. Unfortunately this method of indicating bonding is not followed rigorously, and there are exceptions including, for example, the use of three letters such as BCF, BCV or BCX for gold-aluminium bonding and the use of BF for both types of bonding. Published Data must always be checked for confirmation of these details.

Manufacturers ensure the quality of their semiconductor SMDs by making them to the same consistently high standards as are maintained in the production of the larger and more conventional semiconductor devices. Under any single classification of device type, a manufacturer will offer what is effectively the same chip for both surface mounted encapsulation and leaded encapsulation. Glass passivation is used, particularly for the higher voltage devices, where it is necessary for the intended application.

The rigorousness with which SOT encapsulated devices are tested is indicated by, for example, the test sequence followed by Motorola with their MiniBloc SOT-23 products. Moisture resistance is thoroughly sample-tested with three tests which are conducted where possible to MIL-STD-750. These are known as the 'High Humidity, High Temperature, Reverse Bias Operating Life Test', the 'Moisture Resistance (Cycled Humidity) Test', and the 'Pressure Cooker (High Pressure, High Humidity) Test'. Mechanical strength is also sample-checked thoroughly with three tests. These are the 'Intermittent Operating Life Test' (also to MIL-STD-750), the 'Thermal Shock Test' and the 'Solderability and Soldering Heat Test'. Published failure rates indicate the general sturdiness of the encapsulation, and its comparability with the conventional leaded encapsulations.

4.3.9.1 ZERO-HOUR QUALITY AND FITS

Two ways of measuring quality are frequently used with semiconductors. Zero-hour quality is expressed by several manufacturers in parts per million (ppm) and the subsequent reliability is indicated by a Failure Rate. This Failure Rate is often expressed as Failures In Time Standard. The units are known as FITS, defined as one failure per 10^9 component-hours.

Typical of the quality control systems which the largest manufacturers follow is that of Philips in Europe. Their General Quality Specification has four inspections: process control inspections, group A lot-by-lot acceptance inspections, and group B and C periodic inspections. Group A inspection is in accordance with MIL-STD 105D or IEC 410. Group B and C inspection generally follows CECC 50,000 procedures.

The Parts Per Million quality control system can be implemented only with larger customers because the scale of operation can then support the PPM system adequately. The customer accepts that, with the special packaging systems for surface mounted devices, which include tape and reel, extensive testing of the devices can be executed only by the manufacturer before he packages devices for despatch to the purchaser. A working relationship is built up around the concept that testing of devices belongs almost exclusively to the device manufacturer, and therefore he must accept some responsibility for any failures which may occur later and will be discovered only after the board or equipment has been assembled. Similarly, the equipment manufacturer who is purchasing the devices and assembling them into his boards must accept that he must discuss the assembly methods so that the device supplier is then reasonably confident about sharing responsibility for some board failures. The PPM system with large customers relies on close co-operation and interchange of information.

Acceptance tests are carried out statistically, ensuring that each batch meets Acceptable Quality Levels. These are realistic and practical levels which are set by the device manufacturer taking into account the realistic demands expressed in the market place where he needs to be able to sell the product competitively. In practice, of course, the actual defect levels are much lower than the publicised AQLs, but the leeway or tolerance exists in case it is required. Typical levels are as follows:

Inoperatives, mechanical (visual and marking, including no marking) and electrical;
AQL, individually 0·1%
AQL, combined 0·1%

Marking illegible
AQL, individually 0·25%

Visual appearance defects, and defects in characteristics,
AQL, individually 0·4%
AQL, combined 0·65%

Many of the larger manufacturers are declaring their target as 'Zero Defects'. This target is achievable with good quality control, at some expense, but the already established quality of such products can be seen from the following figures (derived from Philips).

Zero Hour Quality (PPM)

Targets as low as 5 ppm can be established with major equipment manufacturers.

Reliability (FITS)

With the 'stress ratio' defined as $(T_j - 25)/(T_{j(max)} - 25)$ the Failures In Time Standard are as follows.

Professional Equipment:

Stress Ratio ≤   0·3   FITS 10
                 0·3 to 0·5   FITS 20
                 0·5 to 0·7   FITS 60

Consumer Equipment:

Stress Ratio ≤   0·3   FITS 50
                 0·3 to 0·5   FITS 100
                 0·5 to 0·7   FITS 300
                 0·7 to 0·7   FITS 1000

Comparison of these figures with the long established figures for conventional semiconductors will indicate that the reliability of SM semiconductors is virtually identical.

4.4 SURFACE MOUNTED INTEGRATED CIRCUITS

Surface mounted integrated circuits have been available for several years. The earliest surface mounted ICs, developed initially for use in electronic watches, were in the Small Outline Integrated Circuit (SOIC) encapsulations which are now very well-known and are still one of the most widely used encapsulations for surface mounted integrated circuits (SMICs).

Since the SOICs first became available, many additional SM encapsulations have been developed, particularly for the largest chips. Some have become virtually industry standards almost as firmly as the original SOIC outlines. The situation has now been reached where virtually any type of integrated circuit which is available in one of the standard leaded encapsulations can also be made available in an appropriate surface mounted encapsulation. A range of device packages is shown in Figure 4.2(b).

4.4.1 Encapsulations for Surface Mounted ICs

The encapsulations for surface mounted integrated circuits range from the smallest of the Small Outline designs (6-pin), through a number of designs for carrying medium-size chips, up to the more complex chip carriers and pad grid arrays which may have considerably more than one hundred contacts for use with the complex chips. These types of encapsulations are considered as separate groups in the following sections. Some special facets of mounting and soldering large surface mounted ICs are described briefly later.

4.4.1.1 SMALL OUTLINE INTEGRATED CIRCUITS

A view of a typical Small Outline integrated circuit is shown in Figure 4.12.

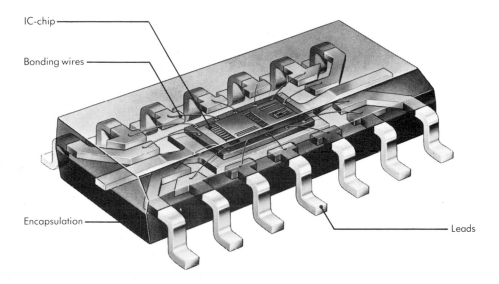

Fig. 4.12 See-through view of Small Outline Integrated Circuit. (Courtesy of Philips)

This is one of a range of which the smallest members were originally developed for use in electronic watches and other miniaturised circuits. Despite the very significant reduction in size compared with conventional integrated circuits, these ICs nevertheless use the same chips and assembly processes as their larger counterparts.

In this Small Outline range of encapsulations, two body-widths are available. One is for the smaller-bodied 6, 8, 10, 14, and 16-lead types, and the other is for the larger-bodied 16, 20, 24, and 28-lead types. In the small-bodied versions, identification of pin 1 is aided by a bevelled edge. On the larger versions there is also a bevelled edge, but, in addition, there is a central notch, similar to that on the conventional DIL packages. Contact spacing is at 1·27 mm intervals for both body sizes. Figure 4.13 shows dimensions of SO8 and SO16 encapsulations.

The type number for an SM device from most manufacturers is generally similar to that of the conventional type, except that a suffix such as D (or sometimes a T) is used for indicating the SM version.

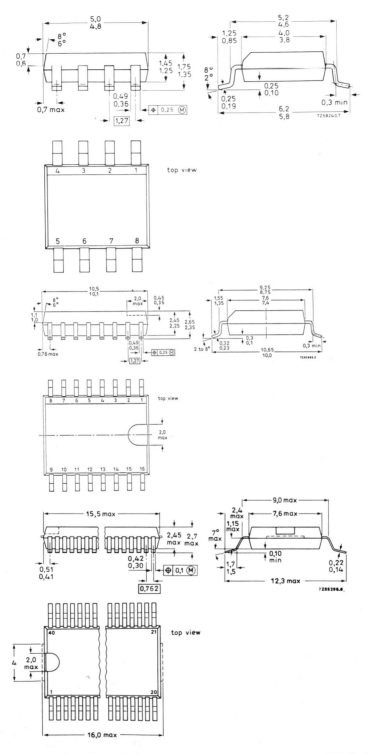

Fig. 4.13 Dimensions of selected SOIC encapsulations. (Courtesy of Philips)

The manufacturer's published Data Sheets must be consulted for all details of type-number coding, and for electrical details but the typical dimensions are shown in Table 4.1, which also includes dimensions of the Very Small Outline (VSO) encapsulations. Because the VSO outline offers a larger number of contacts (with smaller contact spacing), these encapsulations are used for more complex devices.

Table 4.1

		Dimensions of SM Integrated Circuits		
Number of Leads	Width of Plastic (mm, max.)	Length of Plastic (mm, max.)	Max. Width Leads end-end (mm)	Outline SO/VSO
6	4·0	3·75	6·2	SO-6
8	4·0	5·00	6·2	SO-8
10	4·0	6·25	6·2	SO-10
14	4·0	8·75	6·2	SO-14
16	4·0	10·00	6·2	SO-16
16	7·6	10·50	10·65	SO-16L
20	7·6	10·50	10·65	SO-20
24	7·6	15·60	10·65	SO-24
28	7·6	18·10	10·65	SO-28
40	7·6	16·00	12·80	VSO-40
56	11·1	22·00	15·80	VSO-56

The SOICs have 'gull-wing' contacts as can be seen in Figure 4.12. Each individual contact can therefore easily be accessed by probe for both AC and DC testing while the IC is soldered to the substrate.

SOIC encapsulations give all the desirable advantages of conventional ICs, being manufactured in the same way. They have none of the difficulties and risks which are associated with the use of 'naked chips'. Although small, they are robust. Assembly on to boards can easily be effected by a variety of automatic placement machines. Multi-sourcing of the devices offers good commercial protection to the user. With these features, it is understandable and justifiable that the SO package should have become an industry standard for surface mounted ICs.

The level of availability of these devices is indicated by the wide range of products given below. These are extracted merely as examples from the product lists of some of the major manufacturers. Motorola offer, in the 'SOIC Microminiature'[11] outlines, a range of devices including, for example, a selection of operational amplifiers, timing and precision voltage reference ICs, and comparators. SGS offer as part of their 'Micropackage' range operational amplifiers (some with 'Triple Metallisation' for adverse environmental conditions), telecommunication circuits, transistor arrays, various consumer circuits and CMOS HCF4000-Series devices. Philips[12] and their associated companies such as Mullard in the UK and Signetics in the USA offer an extremely extensive range of more than a hundred HEF4000B-Series LOCMOS logic devices, a range of more than fifty 74-Series TTL logic devices, and telephone circuits. Siemens[13] include in their range a variety of operational amplifiers, switching circuits, transistor arrays, amplifiers, and telephone circuits.

Manufacturers are continually extending these ranges, and the existence of multiple-sourcing generally provides adequate commercial safeguard.

For the larger and more complex ICs, the SOIC concept begins to present some disadvantages. Although the Very Small Outline encapsulations extend the SOIC range to 40 contacts (VSO-40) (see Figure 4.13) or 56 contacts (VSO-56), which is sufficient for some complex integrated circuits such as microprocessors, an extension of this design to even larger versions is probably unlikely. This is because of the need to construct unacceptably long internal lead connections and these reduce high frequency performance. Furthermore, an extremely elongated body is inefficient in its use of board space. For these reasons, there is now little motivation to extend the design to larger devices. If the device chip demands a very large pin count, then the styles of encapsulation with leads on all four sides are used.

4.4.1.2   CHIP CARRIERS

During the 1970s several companies introduced an early form of leadless chip carrier. The name 'chip carrier' now applies to a whole range of IC products which have terminations brought out on all four sides of a square encapsulation, or a rectangular encapsulation which is nearly square. The chip carrier can be visualised as the centre portion of the IC with external connections on the four sides rather than in the two rows used by the SO package. The two designs use very similar internal lead-frames but the design of the chip carrier encapsulation facilitates the use of a larger number of pins. Pin-counts range from 16 up to more than 300.

The external dimensions of both the SOIC encapsulation and the chip carrier encapsulation are governed in practice by the number of external contacts required and not by the area of the internal chip. The geometries of the two types of encapsulation mean that with higher pin-counts, the SOIC encapsulation unfortunately becomes disproportionately large. For the more complex integrated circuits therefore, the chip carrier encapsulations use substrate space more efficiently and begin to appear increasingly attractive.

The square design of the chip carrier also permits shorter internal connection paths than are generally feasible with the longer design of large SOICs. Shorter internal connections bring advantages such as lower values of inductance, capacitance and resistance.

To create chip carriers which are suited to the varied requirements of different applications, a number of production methods are used (Figure 4.14). The three layer co-fired method which utilises a flat solder-sealed lid is used for the most demanding applications in military and civil designs where cost is not the ultimate consideration. Less expensive alternative versions of the chip carrier encapsulation use either a single cup-shaped epoxy-sealed lid, or a plastic potting material. Typical outlines are shown in Figure 4.15.

Contact spacing of 0·05 inch is shown in Figure 4.15. The JEDEC standards for 0·05 inch pitch, currently the most popular spacing, were approved in 1980 and have since been revised. Proposals also exist for 0·04, 0·025 and 0·02 inch. There is already a range of standardised packages with overall outline dimensions ranging from 0·235 to 2·05 inch for both the 0·04 and 0·05 inch terminal pitches.

The main disadvantage of the larger leadless chip carriers is the difficulty of matching their coefficient of thermal expansion to that of the substrate. This

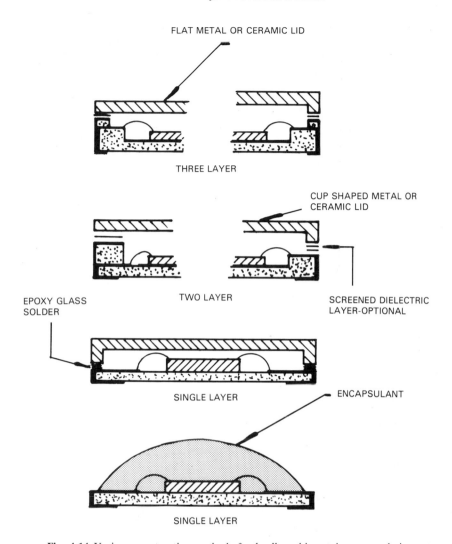

Fig. 4.14 Various construction methods for leadless chip carrier encapsulations.

makes them unsuitable for mounting directly on to conventional glass-fibre epoxy printed circuit board. A few special substrates, including those containing Invar sheets, have been constructed to circumvent this difficulty. A more attractive solution to the thermal expansion mismatch was brought about by Texas Instruments who were one of the first companies to propose a 'leaded chip carrier'.[4] The slightly flexible leads provide the compliance which is necessary when devices assembled on PCBs are to be subjected to thermal cycling. These encapsulations were originally called 'Post-moulded Leaded Chip Carriers'. This name was abbreviated to PLCC, but subsequently the same abbreviation has tended to be regarded as meaning 'Plastic Leaded Chip Carrier'. The device is produced in much the same way as a plastic SOIC but it has J-type leads, on all four sides, 'tucked under' the body in a way which is suitable for surface mounting (Figure 4.16). These J-leads occupy minimal space, being almost totally underneath the body of the device.

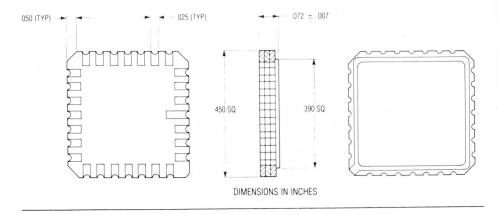

DIMENSIONS IN INCHES

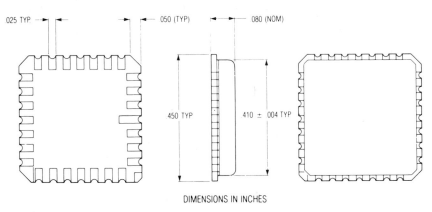

DIMENSIONS IN INCHES

Fig. 4.15 Typical outlines of three-layer and single-layer chip carriers. (Courtesy of Texas Instruments)

An alternative approach to reducing the mechanical stress which is caused by thermal cycling is to mount ceramic and plastic carriers in suitable sockets. Such sockets are widely available and are described briefly in Section 4.4.3.

Although the market uses numerically more SOICS than any other form of surface mounted integrated circuit, the chip carrier is the most popular encapsulation for devices with a pin count exceeding 40 because it offers advantages in space conservation and it has better high frequency performance. For very high pin-counts, say above 164, the leadless grid array or pad array is more likely to be used, and this is described later.

4.4.1.3   THE QUAD FLATPACK

The plastic body of the 'Quad Flatpack' encapsulation is in the shape of a thin square or rectangle and it has compliant leads on all four sides (Figure 4.17). The 'gull-wing' leads are similar to those used in SO packages (Section 4.4.1.1), being compliant and touching the substrate for surface mounted soldering. As with chip carriers, the use of four sides for external connections gives the encapsulation short internal connections. The Quad Flatpack can therefore offer performance comparable to that of chip carriers. The height or thickness of the Quad Flatpack body is less than virtually every other type of large IC

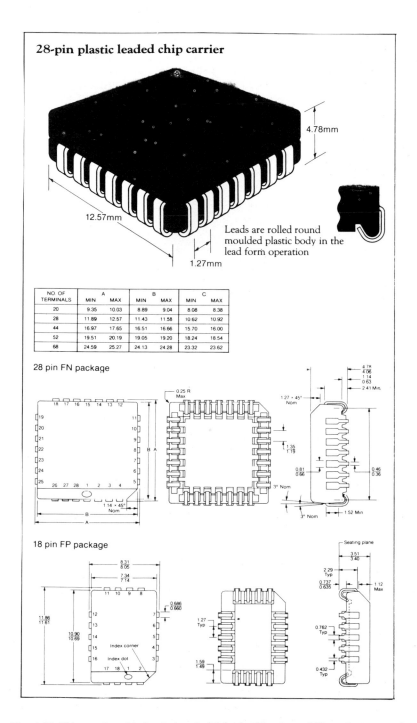

28-pin plastic leaded chip carrier

4.78mm

12.57mm

Leads are rolled round moulded plastic body in the lead form operation

1.27mm

NO. OF TERMINALS	A		B		C	
	MIN	MAX	MIN	MAX	MIN	MAX
20	9.35	10.03	8.89	9.04	8.08	8.38
28	11.89	12.57	11.43	11.58	10.62	10.92
44	16.97	17.65	16.51	16.66	15.70	16.00
52	19.51	20.19	19.05	19.20	18.24	18.54
68	24.59	25.27	24.13	24.28	23.32	23.62

28 pin FN package

18 pin FP package

Fig. 4.16 Chip carrier with 'tucked under' J-leads. (Courtesy of Texas Instruments)

encapsulation, and so these devices are used particularly where there is a need for small dimensions, particularly the small height, and for compliant leads.

Originating in Japan, this encapsulation has been known at various times as the 'Quadpack' and also as the 'Flat Plastic Package' (FPP). Now more widely known as the 'Quad Flatpack', or sometimes simply the 'Flatpack', this encapsulation is still currently used more by Japanese equipment manufacturers than by others. Its characteristics ensure, however, that it will almost certainly continue to gain popularity elsewhere. A number of suppliers, mostly Japanese, offer these packages with lead counts up to 84 or more. Functions include microprocessors, microprocessor and display-driver combinations, and peripheral device controllers.

Fig. 4.17 Typical Quad Flatpack.

4.4.1.4   PAD ARRAY AND LEADLESS GRID ARRAY ENCAPSULATIONS

Complex IC chips can be provided with a very large number of surface mounting contacts by the encapsulation known as the Pad Array or the Leadless Grid Array (LGA). The contacts of a typical array are shown in Figure 4.18.

The surface mounted arrays are derivatives of the conventionally mounted Pin Grid Array (also shown in Figure 4.18 for comparison). This generally has several rows of pins around the periphery of a square or rectangle. Some designs have pins all over the lower surface. Pin-counts in excess of 100 can be achieved in a remarkably small area.

For surface mounting, the Pad Array is manufactured with contact pads arranged in only two rows close to the edge of the device. This design allows substrate conductor tracks of reasonable width to have direct access to the inner conductor pads while still remaining acceptably spaced.

The densely packed pin arrangements of the Pin Grid Arrays is not emulated in the pattern of contacts on the surface mounted arrays for the following reason. The Pin Grid Array is designed on the assumption that multilayer boards will be used, and the pins are so closely placed that connection at different levels is highly desirable. Connection on a single surface would require excessively narrow tracks. Although a similar multilayer approach could be employed with surface mounting encapsulations, much of the advantage of surface mounting would

Fig. 4.18 Typical contact arrangement on Pad Array device, with Pin Grid Array for comparison. (Courtesy of Motorola)

then be lost. To facilitate connection on a single surface therefore, only two rows of pads are used.

4.4.1.5   OTHER SURFACE MOUNTED ENCAPSULATIONS FOR ICs

In the course of the development of surface mounting technology, it is inevitable that many proposed products will be in minority use for a period of time while assessment by customers is deciding their fate. Some designs survive for just a short period and then die, while others thrive. A few of those which do not at once expand into general application are designed strictly for a well-defined minority market. Mainstream IC surface mounting is well established in the encapsulations which have already been mentioned and so only a few of the minority encapsulations are mentioned here.

Encapsulations based on an epoxy substrate constitute an attempt to reconcile coefficients of thermal expansion and are worthy of brief mention. These include the Micropack encapsulation, which is supplied by Siemens[13] and others. The Micropack consists of a standardised substrate $8 \times 4 \cdot 2$ mm in area and $0 \cdot 6$ mm thick, with the chip mounted within a second layer of protective material. Several pin arrangements up to a pin count of 14 are available. Similar devices from other manufacturers include the General Instruments Minipak.[14] The BT EPIC[15] glass epoxy chip carrier is a comparable encapsulation, and there are similar encapsulations produced by Hitachi,[14] Heraeus,[16] and PCI.[14] A specialised market makes use of these types of encapsulation, and they are not strictly in competition with either small chip carriers or SOICs for survival.

The Tape Automated Bonding (TAB) encapsulation cannot strictly be included among surface mounted styles, despite the fact that TAB devices are attached almost directly to the substrate surface and make extraordinarily efficient use of space. In the manufacture of TAB devices the naked chip is attached by thermocompression bonding to small copper pads on a thin flexible polyimide tape. From these contact pads, narrow copper 'leads' on the tape radiate out to a second set of contact pads around the periphery of the chip (similar to the LGA in Figure 4.18), and thence to a third set of contact pads which provide test access points near the edges of the tape. The second set of contact pads can be thermocompression bonded to the substrate during board assembly, and the remainder of the tape and the test leads are then cut free. The total area used by the TAB device is only slightly more than that required for a naked chip, but the specialised automatic thermocompression bonding cannot be mixed easily with the automatic placement and soldering of other surface mounted components. Specialised applications which require large numbers of the same type of chip can nevertheless sometimes be more economical with TAB devices, provided that the investment in the relevant handling machinery is considered.

The descriptions given above are far from exhaustive. No details concerning the current attempts to introduce inexpensive versions of the chip carriers are given. These would include the transfer-moulded Low Cost Chip Carrier from Bell Northern, the Single Level Chip Carrier from Sprague, the 'Tape Pack' from National Semiconductor, and others.

Other forms of high pin-count encapsulations include the Fine Pitch Chip Carriers, the Double-Row Leaded Chip Carrier from Texas Instruments, the Isopak and Versipak based on a Kovar substrate, and the Area Bond TAB which is capable of being adapted for up to 600 connections.

4.4.2   Performance of Surface Mounted ICs

The performance of the compact and lightweight surface mounted ICs is similar to that of the larger versions. In general, the SM versions enable the board area to be reduced to about one third of that achievable with conventional components, and will permit very considerable reductions in mass.

On a standard epoxy PCB the thermal resistance will normally limit the power dissipation to about two thirds of the dissipation of the standard DIL package. On ceramic substrates, however, the thermal conductivity of the substrate can be utilised to restore most of the power dissipation capability. This improvement is achieved by including a heat-sinking compound between the IC and the substrate. For the future, it is likely that an increasing number of routine types will have internal metal heat-spreaders to increase the dissipation limits. Operating temperature ranges are usually given as either 0°C to 70°C or alternatively −40°C to +85°C depending on the requirements.

SO packaged integrated circuits made by Philips use lead frames of copper or alloy 42. For the smaller packages the use of copper lead frames gives a considerable reduction in thermal resistance.[18] Larger packages incorporating the larger sizes of semiconductor dice show a much smaller difference.[1]

4.4.3   Soldering Surface Mounted ICs

Soldering SM integrated circuits can be executed along with all the other SM products by any of the standard immersion or vapour-phase methods. Generally, the limits for ICs are 4 or 5 seconds at 260°C. Although this requirement is considerably more stringent than the requirements imposed by many passive SMDs and by most discrete semiconductor SMDs, it is still sufficient, permitting for example, normal dual-wave immersion soldering. This method of soldering subjects the device to two immersions in rapid succession, each immersion being approximately two seconds.

At the board design stage, adequate size soldering pads should be built on to the board (see Chapter 5). This is a good preparation for soldering and also assists conduction of heat when the device is in operation.

Soldering details for each device are given in published Data Sheets and these should always be consulted. If the upper limits of the soldering conditions for a board are set by the integrated circuits, as may well be the case, care must be taken to produce a set of soldering conditions to satisfy the requirements of all of the devices on the board at the same time.

Post-fabrication servicing or repair of boards generally requires localised hot-air soldering because of the large number of contacts which need to be detached and re-attached. For this operation, stipulated temperature-time limits must always be strictly observed.

Contacts on SMICs generally have the standard finishing coat of Sn-Pb solder, unless specified differently in published data. This finish permits storage for reasonable periods in non-hostile atmospheres and ensures good soldering with any of the conventional hot Sn-Pb soldering techniques.

Not infrequently, conductive adhesive is used for attaching surface mounted integrated circuits to boards which are otherwise already completely finished and are carrying the passive components, connectors, switches and so on. This is desirable in those circumstances where the automatic assembly of larger and more complex devices is not a commercially attractive proposition. For this attachment a silver-filled epoxy is generally used. This permits the assembly of all of the more elaborate encapsulations without raising the temperature to those levels which are required for soldering. The device manufacturer will be fully aware of the occasional need for this approach to the assembly of these larger and more complex devices. He will normally be able to provide details of suitable adhesives, and will be able to state the relevant temperatures and hardening times.

Cleaning of assembled boards which are carrying SMICs can usually be effected by any of the standard cleansing agents such as isopropyl alcohol or Freon. The published Data Sheets should nevertheless always be checked for confirmation of suitability of any proposed cleansing agent.

4.4.3.1   THERMAL CYCLING AND MECHANICAL STRENGTH

When the device and substrate are heated and then subsequently cooled, the dimensional expansion of the board may be different from that of the components. With small components this difference generally causes little or no difficulty but with larger components significant stresses can be set up. Extensive thermal cycling can therefore eventually lead to deterioration and perhaps fracture of joints.

To minimise the risk of difficulties caused by expansion when large components are to be installed on the board, there is a preference for board materials with a small coefficient of expansion. Invar-based boards and many other equally elaborate approaches are used when the expense is warranted.

A general reaction to the potential difficulty of damage caused by thermal cycling is to attempt to use surface mounted devices which have leads. This trend may seem remarkable in a technology in which the components have frequently been given the misnomer 'leadless devices'. Nevertheless, the compliance of the short flexible leads with surface mounting ends gives a degree of protection which is desirable for the larger devices. The need for an inexpensive way to avoid this thermal fatigue may eventually cause the encapsulation designers deliberately to 'invert' the chip and its supporting connection frame inside the encapsulation. The increased length of external leads between the body of the encapsulation and the surface mounting contacts would give the required flexibility. The increase in effectiveness of air cooling of the device would counterbalance the reduction in cooling to the substrate.

An alternative approach to thermal cycling, which points tentatively in a new direction at least for complex circuits, is an extremely flexible polyimide-based 'board' called Flexwell from Welwyn Electronics[19]. Some details of this product are given in Chapter 11.

4.4.3.2   SOCKETS FOR INTEGRATED CIRCUITS

Normally, sockets are used with ICs primarily to facilitate removal for substitution. A further incentive to use sockets for the larger integrated circuits is also given by the thermal stresses which can cause damage to fully soldered ICs. Some sockets have their own leads, but in any case the spring contacts of the socket largely remove the worst effects. With expensive ICs such as microprocessors, there can therefore be considerable advantage in using sockets.

Standard sockets are available from manufacturers who specialise in this type of product, but not generally from the chip manufacturers themselves. A typical socket from Burndy which is widely used is shown in Figure 4.19. Another

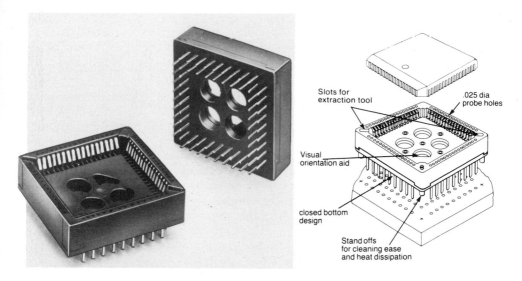

Fig. 4.19 Typical IC socket. (Courtesy of Burndy)

version from AMP, which includes a retaining clip for the chip carrier, is shown in Figure 4.20. The 3M QUIP and the Zero Insertion Force (ZIF) sockets also have their own protective covers. The Advance Interconnections socket for integrated circuits, which has already been mentioned in Section 3.17.3 in Chapter 3, permits the construction of custom-designed sockets.

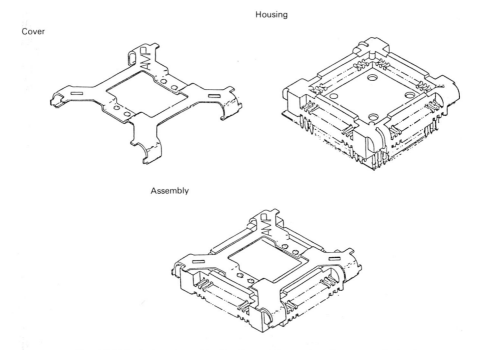

Fig. 4.20 Typical socket with clip above chip carrier. (Courtesy of AMP)

4.4.4 Packaging of Surface Mounted ICs

For the smaller SMIC encapsulations tape and reel packaging is used for delivery to automatic placement machines. Integrated circuit outlines demand tapes larger than the 'Super-8' tape. The 12 mm tapes, 16 mm tapes and 24 mm tapes are frequently used, and larger tapes are being considered by some manufacturers. These are supplied on reels which follow the pattern of those used for smaller devices in their 8 mm tapes. Some manufacturers permit the customer to choose the orientation of the devices within the tape. All MOS devices are delivered in an antistatic tape.

Bulk packaging of large complex devices can be in boxes, with antistatic protection where necessary. Tubes and 'rails' are also sometimes used for those devices where the general shape of the outline (like conventional DIL devices) permits the device to slide in the packaging without interference between the encapsulation and the package.

Quad Flatpacks with leads on all four sides do not slide against one another without causing damage to their leads and hence tubes are not a suitable form of packaging. These devices are generally supplied in 'waffle' packs which are larger versions of the matrix type of packaging used for 'naked' ICs.

4.4.5 Quality and Reliability of SMICs

The manufacturers of surface mounted ICs ensure the quality of their products by employing tried-and-tested procedures derived from the routine manufacture of other devices.

The quality control system used by Philips[20] which is set out below exemplifies the stringent precautions taken by some of the major manufacturers. The checks ensure both early conformity to defined quality requirements and long-term reliability. They are described here both to indicate the quality of these SMICs and also to show the type of approach which ought to be expected. Their test methods during production are generally in accordance with the IEC68-2 series or MIL-STD-883B. Information thus gathered is used both for maintaining optimum process conditions and also for establishing improvements in product design where appropriate.

Tests are performed at combined Absolute Maximum Ratings on all parameters, as defined in IEC Publication No. 134, and 'failures' are complete failures as defined in IEC Publication No. 271.

The quality of the Philips ICs can be seen from the following figures, which are derived from these tests.

Environmental Tests

Combined results of tests on both bipolar and CMOS types.

(a) Solderability after 16 h at 150°C
      sample 500   failures 0
(b) Temperature cycling (up to 200 cycles)
      sample 1942   failures 0
(c) Combined temperature treatments
      sample 406   failures 0
(d) Lead integrity
      sample 350   failures 2

Endurance Tests

Endurance of specified conditions for 1000 hours; bipolar types only.

(a)	Temperature-humidity-bias (85°C, 85%, 15 V) device-hours 60,000	failures 0
(b)	Static life device-hours 28,000	failures 0
(c)	Storage at $-65°C$ device-hours 73,000	failures 0
(d)	Storage at $+150°C$ device-hours 74,000	failures 0

Endurance of specified conditions for 1000 hours; CMOS types only.

(a)	Temperature-humidity-bias (85°C, 85% RH, 15 V) device-hours 243,000	failures 2
(b)	Static life device-hours 368,000	failures 2
(c)	Storage at $-65°C$ device-hours 125,000	failures 0
(d)	Storage at $+150°C$ device-hours 125,000	failures 0

Reliability

Failure Rates (FR) are given at 60% confidence level for the following tests.

(a) Static life at $+125°C$
     FR—$7 \cdot 8 \times 10^{-6}$/h
(b) Storage at $-65°C$
     FR—$4 \cdot 6 \times 10^{-6}$/h
(c) Storage at $+150°C$
     FR—$4 \cdot 6 \times 10^{-6}$/h

The latest figures should be obtained and used as a true basis for comparison, but the general high standards which are reached can be discerned from the data given above. Signetics announced in 1985, followed by Philips in 1986, a zero defects warranty for integrated circuits. Customers can return the entire batch of devices for re-screening or replacement if one defect is found. This policy reflects the current quality situation for ICs and will no doubt be followed by other leading manufacturers.

4.5 SURFACE MOUNTED SEMICONDUCTOR LEDS

Any survey of currently available semiconductor SMDs would be incomplete without some reference to light emitting diodes (LEDs). Telefunken Electronic[21] and Siemens AG[22] produce surface mounting LEDs in the standard SOT-23 outline. The body of the encapsulation consists entirely of a translucent material which totally encloses both the LED chip and all of its internal connections. Externally the device has the standard three SOT-23 contacts. When voltage is applied, the complete 'block' appears to glow. Emission of light in all directions permits viewing from any angle. Versions can be obtained for a range of colour outputs including super-red, green, and yellow. A two-colour (super-red and green) version is also available. Output luminous intensity ranges from about 0·5 millicandelas in red to about 1·5 millicandelas in green. These devices lend themselves directly to automatic handling by the same placement machines as are used for other SOT-23 devices. Delivery can be in 8 mm tape and reel for automatic placement.

Hero Electronics and Elcos GmbH provide versions of the surface mounted LED which are constructed on a ceramic base. In this type of device, known as the Cerled[23], the LED chip is mounted on the top surface of a ceramic block which has dimensions close to the size of a standard SOT-23 device. The diode chip is protected by a clear lens attached to the top of the block, permitting 180° viewing. Cerleds in the CR10 range from Hero Electronics are built on a block 3·2 × 1·27 mm with soldering contact areas formed at the ends (Figure 4.21). Colours supplied include green (with peak at 560 nm), yellow-green or lemon

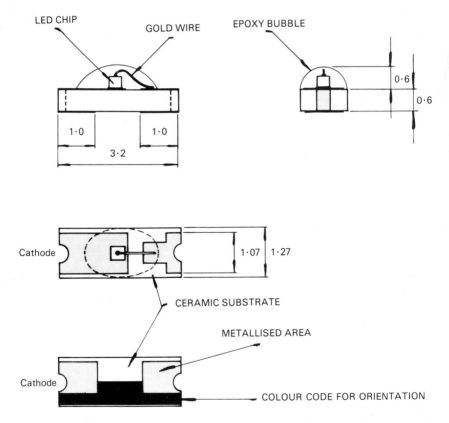

Fig. 4.21 Example of surface mounted semiconductor LED. (Courtesy of Hero Electronics)

(568 nm), yellow (590 nm), orange-red (635 nm), and infra-red (950 nm). Power dissipation is about 180 mW at 25°C, derated linearly to about 130°C. Normal operation is achieved at voltages between 2·1 and 2·5 V. A slightly different design from Elcos GmbH has the solderable contact along the complete side of the ceramic block. Although the body shape of the Cerled is generally very similar

to the SOT-23 outline, the clear plastic lens above the block may inhibit the use of automatic handling. Placement machines may require some adaptation from SOT-23 procedures. Soldering at SM temperatures is permissible. Packaging is in 8 mm tape and reel.

Isocom supply opto-couplers in a surface mountable execution offering a single channel microcoupler on a ceramic substrate. These devices provide a LED output drive current of 20mA.

4.6 AVAILABILITY OF SEMICONDUCTOR SMDS

The foregoing descriptions of surface mounted encapsulations give an indication of the wide range of semiconductor devices which can be obtained. No attempt has been made to list individual types which are available in any given encapsulation. To do this would entail the inclusion of lengthy type-number lists which would inevitably become rapidly out-dated. Similarly, it has not been possible to include direct reference to every individual manufacturer. The majority of manufacturers of conventional devices also now provide semiconductor SMDs and should be approached for information relating to their SM products. The intending user should therefore request current commercial availability information from the appropriate manufacturers or suppliers, and should also request technical data for comparisons between types. Generally, such information is readily obtainable.

REFERENCES

1 'Discrete Semiconductor Chips', Short Form Catalogue CN-164, Sprague Electric Company, Concord, NH, USA.
2 'Microminiature Products', Pamphlet BR127, Motorola Semiconductor Products Inc., Phoenix, Arizona, USA.
3 'Components for Hybrid Microcircuits', Brochure 9398 3036 0011, NV Philips Gloeilampenfabrieken, Eindhoven, The Netherlands.
4 SOT23 Transistor and Diodes Selection Guide, Ferranti Electronics Ltd, Oldham, England.
5 'Diodes', Brochure 9398 323 50011, Philips Export BV, Eindhoven, The Netherlands.
6 'Surface Mounted Semiconductors', Short Form Guide, Mullard Ltd, London, England (1985).
7 'Transistors and Diodes for Surface Mounting', (SMD) 6200 176 1E, ITT Semiconductors, Lawrence, Mass., USA.
8 Superectifier Data Sheets JF 8430 and 8414, General Instrument Corp., Hicksville, NY, USA.
9 E line Data Sheet TD/D 735, Ferranti Electronics Ltd, Oldham, England.
10 'European Type Designation Code System for Electronic Components', Booklet D15, Pro Electron, Bruxelles, Belgium.
11 'Motorola Presents the SOIC Microminiature Package', Motorola Linear Integrated Circuit Newletter, No. 052/81, Motorola Semiconductor Products Inc., Phoenix, Arizona, USA.
12 'SO Miniature IC Packages', Technical Publication No. 085, Philips Export BV, Eindhoven, The Netherlands.
13 'Components for Surface Mounting', Short Form Catalogue, No. B/2943-101, Siemens AG, München 80, Germany.
14 'Surface Mounted Semiconductor Packaging', *New Electronics*, p. 32, May (1983).
15 PCB Industry Information Service, Technology Report, BPA Information Services, Dorking, Surrey, England (1984).
16 'BT Develops Chip Carriers', *New Electronics*, p. 7, July (1983).
17 Herastrat brochure, No. PK-V10, VC Heraeus GmbH, Hanau, W. Germany.
18 'Thermal Resistance of SO Packaged ICs', Technical Publication, No. 183, NV Philips Gloeilampenfabrieken, Eindhoven, The Netherlands.
19 Kirby, P., 'A Flexible Approach to Chip Carrier Mounting', *Electronic Engineering*, Reprinted by Welwyn Electronics Ltd, Bedlington, Northumberland, England.

20 'Surface Mounted Devices', Integrated Circuits, No. 9398 316 40011, Philips Export BV, Eindhoven, The Netherlands.
21 'New Opto Products', brochure AEG (UK) Ltd, Slough, Berkshire, England.
22 Infoservice, No. 1138413, Siemens AG, Furth, W. Germany.
23 'Report of a Novel Opto-Electronic Component—The CERLED', Hero Electronics Ltd, Ampthill, Beds, England.

Chapter 5

DESIGN OF THE CIRCUIT LAYOUT

G. A. WILLARD, J. F. PAWLING and K. R. STONE*

Mullard Mitcham, Mitcham, Surrey, England
*Formerly of Mullard Mitcham, Mitcham, Surrey, England

5.1 INTRODUCTION

The earlier chapters (3 and 4) have shown the range of active and passive component types available in Surface Mounting form. It is evident from this information that a complete range of devices required to manufacture circuits is now available. Active components ranging from single diodes, through transistors and logic integrated circuits to complex microprocessors are obtainable in a range of different packages suitable for Surface Mounting. They can be used together with the supporting cast of passive devices to enable circuit functions and even complete systems to be assembled in Surface Mounted form.

Substrates (Chapter 2) are also available in a very wide range of materials to suit different application requirements and environments, ranging from simple single-sided 'paper' printed circuit boards to multilayers, flexibles, metal cored and insulated metal substrates. Traditional thin film and thick film substrates of glass and alumina are also suitable candidates for the use of SMDs; indeed the earliest use of SMDs was to provide the active devices and high value capacitors required to enhance these additive film techniques used to produce interconnections, resistors and low value capacitances. The range of sophistication of substrates and hence the range of cost is very wide and the designer must weigh up a large number of conflicting requirements before making his or her final choice. Line widths and spacings can vary widely from 'state of the art' close spacings and fine line widths (typically 0·005 in.), to the simplest and cheapest larger dimensions which, however, offer less dramatic reductions in the area taken by the substrate.

5.2 DESIGN OPTIONS

The design engineer, with Surface Mounting Technology available to him, has an extra degree of freedom over his colleague who is limited to leaded devices only. He has the choice of using this extra freedom, or not, depending on the needs of his design in terms of size, price, etc.

Some technologies developed in recent times have been limited in their range of applications due to restrictions of power handling, or size of suitable substrates.

This size restriction led to thick film being basically limited to smaller sub-systems or parts of circuit functions and the need then to mount these DIL or SIL sub-assemblies onto some other 'mother' board to form the complete system. Surface Mounting Assembly Technology does not suffer from this sort of restriction; for example, SMD resistors are limited to about 250 mW dissipation, but they need not be the only type used in an assembly. SMDs may be used, together with existing components, on and through a common substrate, thus forming a 'Mixprint' assembly on a printed circuit board (Figure 5.1). This combination is

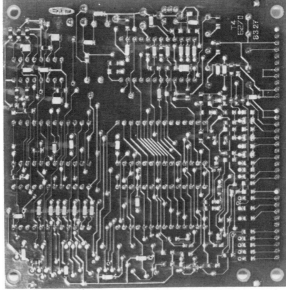

Fig. 5.1 Typical 'Mixprint' board assembly using both leaded and surface mounted components. (Courtesy of Mullard Ltd)

capable of producing a complete circuit or system as a single board, the large devices and power components being mounted conventionally alongside the smaller SMDs.

The Mixprint assembly may use a single-sided paper-based printed circuit board with leaded components on the top surface and SMDs on the bottom (soldered) surface to achieve a very low cost assembly, but with a relatively small size reduction compared to its leaded counterpart. Progression through double-sided through-hole-plated boards to multilayers and on to fine line, buried 'via' multilayer boards gives increasing size reduction but with a cost penalty because of the increasing complexity of the substrate system. This increasing complexity, however, offers greater packing density for the components with the chance to convert more devices to Surface Mount types when they are no longer required to perform the function of track jumpers.

The Surface Mounted components may also be placed on both the top and bottom surface of the substrate, and, in the ultimate situation with a buried via multilayer system, both surfaces of the substrate may be completely covered with components, the interconnection being performed in the interior of the substrate between the two layers of components mounted on its surfaces. The top surface is usually taken to be that which in a mixprint or conventional assembly includes the bodies of leaded components. SMDs used in this type of assembly have advantages over their leaded counter-parts, not only in reduction in size, but also because their leads do not penetrate the board and hence do not take up space on the layers to which they do not connect. A leaded device will automatically connect to all layers, but this may be a disadvantage when very dense packaging is required.

Some of the possible PCB variations with Surface Mounted components are summarised with the aid of Figure 5.2. This shows assemblies exclusively using

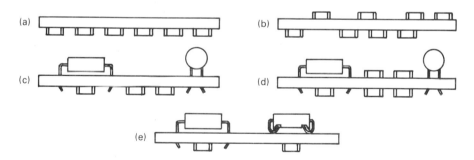

Fig. 5.2 Variations of PCB assemblies. (a) and (b) SMD only; (c), (d) and (e) mixprint.

SMDs on one side, or both, SMDs mixed with leaded components and finally a mixprint requiring both wave and reflow soldering. (a) and (c) can be accommodated with a single layer board while the others require two or more layers of interconnect.

The use of these considerably more complex substrate boards to achieve greater size reduction normally precludes the use of large numbers of leaded devices, especially large ones, as the cost of the substrate itself means that its surface must be used as efficiently as possible. This normally implies that the board only performs a part of the circuit or system requirement and becomes a 'daughter' board to be mounted in SIL or DIL form onto a 'mother' board. In this way the

SMAs mirror the thick or thin film sub-circuit currently used in this form in complete systems.

In certain environments the temperature range encountered and the packaging of the devices which need to be used, for example, leadless ceramic chip carriers, mean that some form of thermal matching of the devices and the substrates must be achieved to enable a reliable product to be manufactured. Alumina substrates or cored boards of various types may be used in these cases and, although multilayer interconnects become more difficult, very dense packaging can be achieved with certain circuits, e.g., memories.

The design engineer must weigh up all the requirements of the application and decide on which type of assembly is required. SMA can achieve a size reduction with a Mixprint assembly on a single board at low cost. Greater size reduction normally calls for increasing complexity of the substrate or the use of mother/daughter board arrangements to form an effectively three-dimensional assembly. Examples of various design options are discussed further in Chapter 9.

5.3 DESIGN RESTRICTIONS

A number of other factors may affect the choice of assembly type:

5.3.1 CAD System

The available design system (CAD) may not be able to handle a mixture of leaded and surface mounted components on both sides of the board and hence partitioning of the circuit to SMA and leaded portions may have to be used. The problems of CAD systems relative to SMA are dealt with in more detail later in this chapter (Section 5.5).

5.3.2 Production Machinery

The designer who designs a board or assembly without due regard to the available equipment for manufacturing the assembly is courting disaster and this is even more important in SMA when high speed machinery is to be used.

The soldering systems used for SMDs vary depending on which surface of the board the devices are mounted on, and the design rules for these differ in detail. The designer must consider the systems to be used at the board design stage of the project. The use of certain types of components may, in fact, preclude some methods of soldering altogether.

Testing of the final board assembly is nearly always required and the methods used will impose some constraints on the designer which must be taken into account at an early stage in order to avoid excessive time and hence cost of testing.

These three constraints—placement, soldering, and testing—are now discussed in greater detail.

5.3.2.1 CONSTRAINTS IMPOSED BY PLACEMENT MACHINERY

Surface Mounted Assembly is a machine-based process, hand assembly of the very small, often unmarked devices being an almost impossible task when large numbers of SMDs and large numbers of substrates are involved. The speed of placement achieved in production is very dependent on the board design, and the designer must be aware of the requirements of the production machinery from the outset. The reduction of machine movement between placements will not only increase the machine speed and reduce the production time and cost, but will also reduce the wear on the machine parts and increase its life.

Surface Mounting Placement machines cannot hold an unlimited number of different components and the selection of types and values used in the circuit design should take account of this restriction. The E12 range of 5% resistors alone has 85 different values to cover the range from 1 ohm to 10 Mohms. A typical $120,000 machine can handle a maximum of 120 different components (Figure 5.3). Repeated placement of the same component value from a single

Fig. 5.3 Component reel holders on a placement machine.
(Courtesy of Dynapert Precima)

feeder onto different board locations is possible on nearly all machines but on some (e.g., Philips MCM range) this is not a very efficient mode of operation. In this case a large number of Surface Mounted Components per board is best handled by multiple placement of the same complete set of devices or by a second pass through the machine with a different placement programme and a new set of components. (This machine uses 'simultaneous/sequential placement' where 32 components per head are picked up simultaneously and then placed sequentially. Multiple placement from a single reel is therefore an inefficient mode of operation. See Chapter 6 for further details of machines.)

The available machinery may also place further restrictions on the designer in terms of the optimum mixture of components on the two surfaces of the substrate. If two machines are available, for example, with the following capabilities:

(i) fast machine, limited to 8 mm taped devices only;
(ii) slow machine, capable of handling all devices up to large ICs,

then the designer should not mix small 8 mm taped devices with larger ones on both sides of the board as this could mean both sides of the assembly being processed by the two machines. A design which puts all the large devices on the top surface and all the small 8 mm taped devices on the underside would give considerably more efficient use of the equipment.

5.3.2.2 THE INFLUENCE OF THE SOLDERING METHOD

When a range of soldering methods is available, the design of the substrate can determine the particular method to be used. Alternatively, the available soldering method may dictate to the designer and place many constraints on him, not only restricting his layout design freedom, but also narrowing his choice of components.

There are two main methods of soldering SMDs—wave soldering and reflow soldering (Figure 5.4). Wave soldering is ideal for mixed print assemblies where

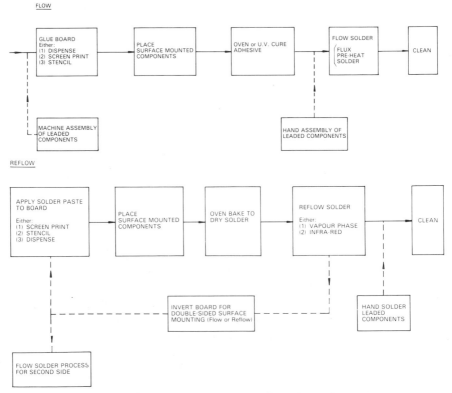

Fig. 5.4 Schematic representation of basic flow and reflow soldering processes.

there is a large proportion of leaded devices and the density of SMDs is relatively low. It is also the cheaper of the two methods and is a process familiar to PCB assemblers. Reflow soldering is preferred for assemblies with SMDs on both sides and for the more densely packed assemblies. This can be the more expensive process, both in terms of capital outlay and running cost. Both of these processes are described in greater detail in Chapter 7. The two processes have very different requirements which have led to two distinct sets of design rules being developed by SMD users. These sets of design rules are not interchangeable so the decision on which soldering method is to be used must be made before the PCB is laid out. Wave soldering can be split into two categories—jet soldering and dual wave soldering, which are more fully described in Chapter 7. In these processes the components are glued to the PCB and the assembly is passed through a jet of molten solder that meets the component terminations and pads and solidifies to form solder joints.

It is necessary to retain the SMDs in position until they have been soldered. This can be achieved either by the solder paste used with reflow soldering or by means of a non-conducting adhesive in the case of wave soldering. The attachment of components using adhesive is one of the main problem areas with surface mounting. This is applied to the PCB between the component pads so that the centre of the component is glued to the board. It is important not to contaminate the solder pads or component terminations. It is equally important to apply sufficient adhesive to prevent the component moving before it has cured. This calls for a spot of glue with minimum diameter and maximum height. To reduce the distance between the underside of the component and the top of the PCB, a dummy track is placed under the centre of the component (Figure 5.5).

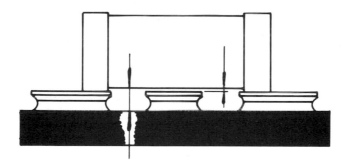

Fig. 5.5   Use of dummy track under component to reduce height of glue dot required.

Once the components have been stuck down, they will not move during soldering, which has the advantage of allaying some of the problems found in reflow soldering as described in the next section, but does not have the advantage of self alignment.

When using wave soldering, the joint is made between the end face of the component termination and the solder pad (Figure 5.6a). The underside of the component termination plays little part in the bond. For this reason, the solder pads must extend beyond the end of the component and be large enough to pick up sufficient solder from the wave to make a satisfactory joint. Careful consideration must also be given to the direction in which the assembly is to be soldered to ensure that components are not shadowed from the solder wave by

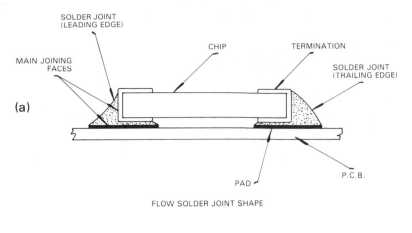

SOLDER JOINT
(LEADING EDGE)

CHIP   TERMINATION

MAIN JOINING
FACES

SOLDER JOINT
(TRAILING EDGE)

(a)

PAD

P.C.B.

FLOW SOLDER JOINT SHAPE

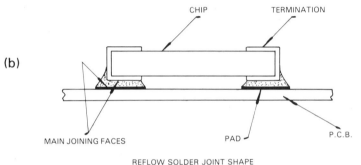

CHIP   TERMINATION

(b)

MAIN JOINING FACES   PAD   P.C.B.

REFLOW SOLDER JOINT SHAPE

Fig. 5.6 Basic flow solder and reflow solder joint shape as seen in a sectional view.

other components. The distances between components must be such that bridging does not occur, which will depend on the type and setting of the solder machine being used.

The reflow soldering process is more fully described in Chapter 7, but basically the solder is applied in its solid form to the areas where the joints are required and then it is melted or reflowed to produce soldered joints. The joint is formed between the underside of the component termination and the substrate, the side or end of the termination playing a relatively minor part in the joint (Figure 5.6b). The solder is placed only in the areas where it is needed and in precise quantities, thus reducing the possibility of solder bridging. This enables much denser assemblies to be produced. The process is not directional so there is no need to lay out the components in rows as in wave solder, nor is there any need to consider shadowing effects as these do not occur. The solder pads can be made smaller because there is no longer the requirement to protrude past the end of the component to 'pick-up' solder from the passing wave. It is, however, advisable to have some protrusion of the pad past the edges of the component for several reasons. The larger the pads, the less accurately the components need to be placed. Larger pads enable more solder to be applied, particularly when solder paste is screen printed where there is a limit to the thickness of solder that can be applied in one pass. It is also advisable to be able to see the solder fillet on the end of a component to inspect the joint. The spacing between components is largely dependent on the accuracy of the machine placing the components, but it should be kept in mind that the components tend to align themselves during reflow. When the solder meets the ends of the component, surface tension effects tend to pull the components towards the centre of the pads. This can also be detrimental and the same effects can pull the components out of line, particularly if pads of different size are used at opposite ends of a component. The pads on both ends of components should be the same size and shape—plated-through holes and large areas of land should be separated from the pads by means of a short length of track covered in solder resist. This is to ensure that one end of the component does not solder first causing the 'Tombstone' effect as described in Chapter 7, and also to ensure that the solder is not drained away from the joint by the plated-through holes or large areas of land.

5.3.2.3 INFLUENCE OF TESTING REQUIREMENTS

One of the main problems of testing SMDs at the component level, when soldered onto the substrate, is that there is no easy and reliable way to access the ends of the components. Gull wing leaded devices (SO ICs, SOT23 semiconductors, etc.) may have their leads bent or pushed momentarily into contact with the PCB by the force exerted by the test probe, when they were not actually in contact with the substrate as assembled. The test probes may also be deflected by the steeply sloping parts of the SMD solder joints causing damage to the devices or PCB tracks. The tolerances of the test fixture PCB alignment holes and device placement positions may be such that the probes may miss the solder pad or hit the component body itself[1].

The designer should consider the requirements of testing at an early stage and design suitable pad extensions or dedicated test pads into his substrate layout to suit the test equipment and test probe systems intended to be used (illustrated in Chapter 8). On assemblies which are completely surface mounted, this may increase the size of substrate required to such an extent that the SMA shows very little reduction in size relative to a leaded assembly. Mixprint assemblies may overcome some of the problems if the leads of the non-SMDs are probed and testing is not taken down to single component levels. Repair of these boards would generally require replacement of a number of SMDs, rather than one device when a fault is found, but this may be an acceptable price to pay for the size saving achieved by omitting the test pads.

5.4 GENERAL DESIGN RULES

There is no one set of component footprints (Figure 5.7) that will work for all situations due to differences in processes, equipment and component packages. There are, however, guidelines which can be used to generate a set of footprints for a particular process and application.

5.4.1 Guidelines

The general guidelines for chip resistors, capacitors and SOT23 semiconductors are now described and these can be adapted to derive pad configurations for other component package types.

The main influence on pad sizes and shapes is the overall placement accuracy of the process; this includes positional accuracy determined by the placement machine, the accuracy of the pattern on the substrate and the dimensional

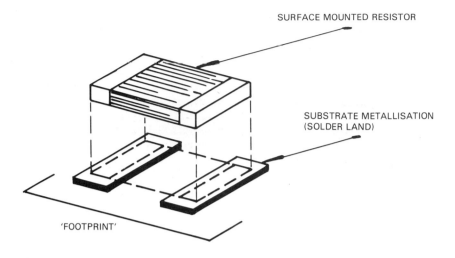

Fig. 5.7 'Footprint' of SMD.

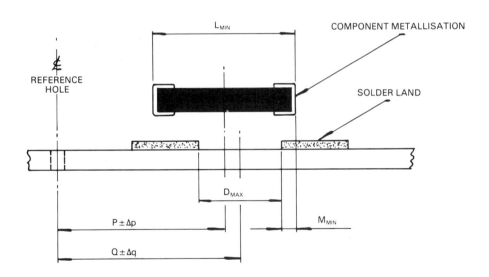

L_{MIN} = MINIMUM LENGTH OF COMPONENT

D_{MAX} = MAXIMUM DISTANCE BETWEEN SOLDERLANDS

M_{MIN} = MINIMUM OVERLAP OF COMPONENT METALLISATION AND SOLDERLAND

P = NOMINAL POSITION OF COMPONENT (TOLERANCE Δp)

Q = NOMINAL POSITION OF PATTERN (TOLERANCE Δq)

$$D_{MAX} = L_{MIN} - 2M - 2\Delta p - 2\Delta q$$

Fig. 5.8 Calculation of space between solder pads.

tolerances of the components themselves. The worst possible situation where all the errors work against each other must be considered to ensure a reliable product.

The width of the pads is usually the same as the nominal width of the component termination; this can be increased for components with very small terminations, such as SOT23 semiconductors, to ease placement problems.

The length of the pad is determined by the spacing between the pads and the amount of land required to extend beyond the component. Usually the maximum possible distance is required between pads to allow tracks to be run beneath the component. This distance can be calculated by assuming the shortest length component, subtract the required amount of overlap of both the terminations on the pads (usually 0·1 mm), and then subtract maximum errors in positional and substrate accuracy, i.e., taking the worst case where the positional error is in one direction and the substrate error is in the other direction (see Figure 5.8). The lengths of the pads are then determined by considering the longest component at its most misplaced position and then adding sufficient length for satisfactory soldering, (e.g., 0·4 mm for dual wave soldering). This will give a typical footprint for, say, a 1206 component as shown in Figure 5.9.

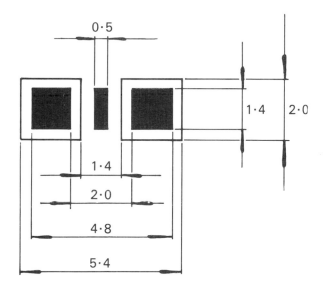

Fig. 5.9 Typical footprint for a 1206 resistor for use with flow soldering.

The spacing between components is calculated in a similar manner but an additional factor is included which is the minimum distance required to avoid bridges and short circuits; this is an empirical figure which must be found by experimentation. The factors to be taken into account in arriving at these empirical figures are as shown in Figure 5.10. Surface Mounted devices may be assembled relative to one another in three different ways. The clearances required between the devices depend on the way in which they are mounted, their size, and the soldering method to be used. The separations as recommended by Philips for their components with dual wave soldering are shown in Figure 5.11. If reflow soldering is employed, these empirical figures can be reduced considerably, due to the reduction in the likelihood of bridging.

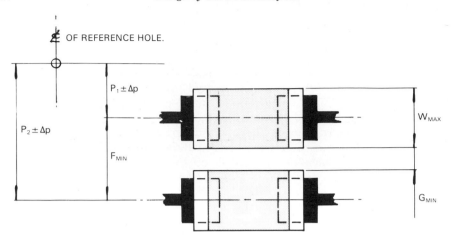

W$_{MAX}$ = MAXIMUM WIDTH OF COMPONENT

G$_{MIN}$ = MINIMUM PERMISSIBLE GAP BETWEEN COMPONENTS

F$_{MIN}$ = MINIMUM PITCH

P$_1$ = NOMINAL POSITION OF COMPONENT 1. (TOLERANCE Δp)

P$_2$ = NOMINAL POSITION OF COMPONENT 2. (TOLERANCE Δp)

$$F_{MIN} = W_{MAX} + 2\Delta p + G_{MIN}$$

Fig. 5.10 Factors affecting the minimum spacing between components.

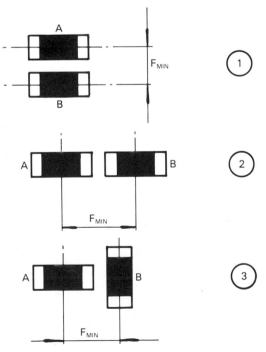

Fig. 5.11 Three methods of placing adjacent components. Dimensions given on next page.
(Courtesy of Philips)

DIMENSIONS:

COMPONENT	1	2	3
R/C 1206/R/C 1206	2·8	5·4	4·1
R/C 1206/CO 805	2·8	4·8	4·1
CO 805/CO 805	2·6	4·2	3·5

Fig. 5.11 Three methods of placing adjacent components.
(Courtesy of Philips)

A series of footprints for some of the many different surface mounted components are given at the end of this chapter for the flow soldering and the reflow soldering processes (Figures 5.13–5.19). The basic pad shapes and sizes are defined for a particular CAD system and plotting equipment and minor variations may work equally well. The footprints shown have been successfully used in a range of Mixprint and full SMD assemblies designed and made by Mullard Limited.

5.5 COMPUTER-AIDED DESIGN OF THE SUBSTRATE LAYOUT

5.5.1 General Requirements

The objectives of a good layout design are to position the components so that optimum use is made of the substrate area and subsequent manufacture of the electronic assembly is achieved with high quality and efficiency. It is possible to produce artwork manually for printed circuit boards or hybrid circuits using colour pencils and paper for the initial design and black tape stuck onto clear plastic film for the final artwork. However, one of the aims of SMA is nearly always miniaturisation. This requires close positioning of components, small conductor pad sizes, narrow conductor widths and spacings and these, in turn, demand more accurate artwork. Greater accuracy can be obtained by producing the master artwork several times larger than the eventual layout, and then photographically reducing to the required final dimensions, but far greater accuracy is achieved by digitisation.

Digitisers first began to be popular in the 1960s and employed the concept that all dimensions and positions are referred to a standard pitch X and Y grid. A computer will correctly interpret the approximate positioning by a designer of, for example, the centre of a component fixing hole, and with great accuracy describe the intended position by its X and Y co-ordinates on the grid. Digitisers thus enable accurate artwork to be produced, but do not assist the engineer in creating his design.

Computer Aided Design equipment for printed circuit boards became generally available in the early 1970s, although simple routing algorithms first appeared as early as 1961[2]. They helped the designer to produce layouts more quickly but, more importantly, as circuits became more complicated, they made what was becoming impossible possible again.

5.5.2 PCBs with Inserted Components

The first decade of computer-aided design equipments was aimed at the printed circuit board with inserted components. More specifically the majority were aimed at digital or logic circuits using significant numbers of dual-in-line integrated circuits with two or more layers of conductors. The single-sided analogue board neither required CAD to the same extent nor was it so easy a task for the applications software writer to offer automatic routines for component placement and conductor routing. These first generation equipments reasonably assumed that all the components were on one side of the substrate contrary to the usual needs of SMA. The following sections will first look at these machines in a little more detail and then consider the special needs of SMA.

5.5.2.1 CAD EQUIPMENT

The CAD hardware will include one or sometimes two cathode ray tube displays—either monochrome or colour. The screen will show graphically the present condition of the layout design and, either on the same screen or a separate one, will supply other data to the designer, such as how many connections are still to be completed, what choices are open to the user and so on.

Information must first be given to the computer. A combination of various methods can be employed including a full qwerty alpha-numeric keyboard, a limited number of keying pads, light pens onto the CRT surface, electronic pens and 'mice' onto a position sensitive working pad area, joysticks and touch sensitive screens. Two further essential elements of any system are sufficient computing power and memory, and adequate applications software.

Some of the earliest CAD equipments included software written by the user. Today's sophisticated systems exclude the 'Do It Yourself' approach as several hundreds of man-years are necessary to write and debug the programmes. Thus, the choice is whose software to buy. In many instances this is linked with the hardware as vendors tend to offer a 'Turnkey' package of terminals, computer and software packages. More recently it has become possible to purchase a more general purpose hardware work station which can run a number of application packages including one for PCB layout design.

When purchasing CAD for the first time a number of decisions need to be taken. The first question is how many work stations are likely to be needed. Care is needed here as CAD installations seem always to attract more work than was originally envisaged. The possibility of shift working needs to be considered and its practicability confirmed or rejected. In practice the computing power of a multi-terminal set-up can be evenly distributed locally to each terminal, can be totally centralised, or can be a compromise between these two extremes. If only one work station is ever likely, the decision is easy. Otherwise considerations include the required speed of response, the desire for an easily managed central data base and how much money is available! The interaction between designer and computer is important. Various vendors offer different tools for this, and personal choice will enter here, although most will agree that graphical input rather than alpha-numeric is preferable. Monochrome displays can be adequate where the number of layers of components and conductors is limited and sufficient differentiation can be achieved with different solid and dotted lines of varying brightness. As the layers increase in number, colour becomes increasingly attractive. Before considering the major software applications of automatic component placement and conductor routing, other facilities worth mentioning are the ability to zoom in and out and pan both vertically and horizontally. These are especially desirable with large area substrates and can be achieved in either software or hardware. The final major technical decision is the degree of software sophistication required.

5.5.2.2 CAD SOFTWARE

The software is clearly dependent on the computing power available. Although CAD software can be run on a personal microcomputer, the routines offered will be limited compared with that achievable on a bigger computer. The more powerful auto-routing and interactive graphics will demand significant computer capacity. When choosing software the user must be sure of his need. If high frequency analogue circuits are the norm, then auto placement is not even necessary, but on the other hand, it can be very helpful in a digital system containing a high proportion of ICs. If the user is a specialised and highly skilled engineer, spending all his time at the work station, then a different approach is desirable from one where an occasional or semi-skilled person is used to operate the computer. Once answers are available to these questions, one can turn one's attention to the sequence of steps that are followed in a typical layout design.

5.5.2.3 STEPS IN A LAYOUT DESIGN

Information must first be given to the computer. This will include the substrate size and details of the circuit to be laid out, such as which components are used and how they are interconnected. Unlike design methods using pencil, paper and rubber, the data cannot be later lost or inadvertently changed, but it must be correctly entered in the first instance. This can be achieved by duplicating the connection data preparation using two different people and checking for discrepancies or using the information first to produce a schematic circuit diagram and have this confirmed by the circuit designer.

A further advantage of CAD is that standard library shapes for component connection pads, and their relative spacings, can usually be stored in the computer's memory and simply called up by the designer, reducing the work load and encouraging a standard approach to the design function. Layout design proceeds by making two types of decision—where to place the components relative to each other and what is the best route for the interconnections between component terminals. The degree of assistance offered to the designer by the computer will depend upon how powerful it is and its application software. Various algorithms are employed to provide automatic placement and routing. With most systems there will come a time, however, when the operator will be required to work interactively with the computer and assist in the design process. This may be because many systems do not include a so-called rip-up approach, which is simply the ability to see that, by changing a previously made decision, greater freedom for further progress can be achieved. Thus the human intervention may be merely to move a component's position or the routing of a conductor and then let the computer try again.

Once the layout design is completed a set of artwork masters can be produced from a photo plotter. Data can be automatically routed from the CAD computer to the plotter via a wire link, or, alternatively, the information can be sent to a remotely sited plotter by any acceptable medium such as punched tape or floppy disc. Output data can also be taken from the CAD computer to control automatically a printed circuit board hole drilling machine, to drive a printer and produce a parts list or a drawing of the final layout, and to provide the component positioning data for computer controlled component insertion and placement machines. It is also possible to supply data for in-circuit component testers.

Much of the above description applies to the layout design of SMA and mixed technology substrates, but was based on machines designed for inserted components on printed circuit boards. The following section considers the specific needs of SMA and mixed technology and the limitations of the earlier CAD systems.

5.5.3   SMA and Mixed Technology Layouts

One advantage of SMA is the ability to place a component on either side of the substrate, particularly when this is a printed circuit board. The ability to choose which side, and in a mixed technology to choose between a chip component or its wire terminated equivalent, offers considerable freedom to the layout designer. An example can be the choice between bridging several conductors with the insulated component body of a wire-ended resistor in one place, or using in another the smallest chip resistor when space is at a premium. These freedoms were not, however, anticipated by earlier CAD systems and in effect make the design problem three-dimensional rather than two-dimensional as with DIP designs.[3] Other limitations, when trying to design an SMA with these earlier machines, are next described.

To use CAD to maximum advantage with components on both sides of the substrate, it is obviously desirable that the user can quickly see on which side any particular component is situated. Often, however, with earlier systems, this was not possible—even when using a colour CRT. This point is illustrated in monochrome in Figure 5.12 which shows first the component outlines of the two sides of a simple SMA, then the two sides combined and, finally, the connections of just one side added. By now the picture on the display screen is not very clear. The designer would like to use different colours for the two sides but, surprisingly, this facility is not always available. Furthermore, some systems do not recognise the possibility of components being placed directly opposite each other on either side; some reject the proposition as a fault and others, while allowing this situation, will not allow the connection pads to apparently clash. This latter feature stems from the application software writer's assumption that a double-sided printed circuit board uses plated-through holes either as solder connections for inserted component leads, or vias taking conductor connections through from one side of the board to the other. Thus, with SMA, it is necessary to define on which conductor layer a connection pad is situated. The pads should not have via holes in them; these should be indicated separately. If leaded components are being inter-mixed with SMA components, they require pads on all the layers that require interconnections.

If two identical SM components are to be mounted on opposite sides of the substrate, then the library shapes will be required to ensure the connection pads are on the correct side and that the pin configuration for an asymmetrically pinned component is correct for that side. In other words, it will be necessary with some systems to define two library shapes of the same component for each side of the substrate, one being the mirror image of the other.

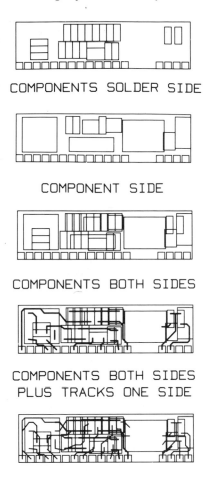

COMPONENTS SOLDER SIDE

COMPONENT SIDE

COMPONENTS BOTH SIDES

COMPONENTS BOTH SIDES
PLUS TRACKS ONE SIDE

Fig. 5.12 The problem facing an SMA layout designer using a monochrome CAD system in determining which components and connections are on which side of the substrate.

The small body sizes of SMDs combined with large substrate areas mean that much larger numbers of components can be placed on one layout. Earlier CAD systems did not anticipate such numbers and the result can be that not all the components can be placed on the standard dimensional grid.

This may lead to other consequential problems such as the situation whereby clearances cannot be automatically checked by the computer. The smaller physical dimensions of the bodies and pin spacings of surface mountable components limit the number of conductor tracks that can be routed between component legs and connection pads. This, in turn, demands a more careful choice when selecting IC gates and ports. There may also be shortcomings with the photoplotters used to produce the layout artwork. They may have difficulty in resolving the finer features of SMD or be unable to satisfy the extended range demanded by a mixed print technology.

The preceding comments are not intended to suggest that SMA layout designs cannot be greatly assisted by earlier CAD systems, but to warn designers of some of the possible problems. More modern, so called 'Grid-less' or 'Grid-free', systems (references 4, 5 or 6) have resolved many of these points but not all designers have access to the latest equipments!

REFERENCES

1 'Probe Problems with SMDs', *Electronic Production*, p. 38, August (1985).
2 'Routing Evolution: Survival of the Fittest', *Electronic Design*, p. 191, 16 May (1985).
3 'Integrated System Accommodates Surface Mounting', *New Electronics*, p. 73, 10 December (1985).
4 'Circuit Board Layout System Runs on CAE Workstation', *Computer Design*, p. 177, June (1985).
5 'Design Automation Tools Aimed at SMD', *Electronic Product Design*, p. 45, December (1985).
6 'Boardmaster—Daisy Systems. CAD and CAE Join Forces to Reduce PCB Design Times', *Electronic Times*, p. 41, 27 June (1985).

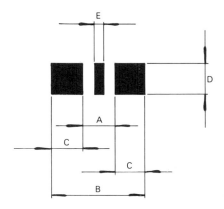

DIMENSIONS:

WAVE SOLDERING					
COMPONENT	A	B	C	D	E
0805 C and R	1·2	3·6	1·2	1·2	0·4
1206 C and R	2·0	4·8	1·4	1·4	0·5
1210 C	2·0	4·8	1·4	1·4	0·5

REFLOW SOLDERING					
COMPONENT	A	B	C	D	E
0805 C and R	1·0	2·7	0·85	1·25	0
1206 C and R	1·95	3·95	1·0	1·6	0
1812 C	2·8	5·2	1·2	3·2	0
2220 C	3·7	6·5	1·4	5·0	0

Fig. 5.13   Footprints for flow and reflow soldering of Surface Mounted capacitors and resistors.

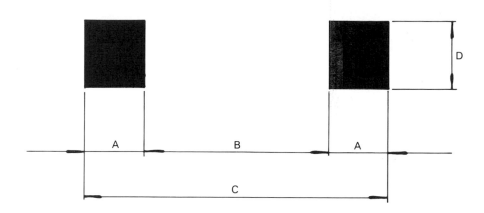

DIMENSIONS:

WAVE SOLDERING				
CASE SIZE	A	B	C	D
1a	2·0	6·0	10·0	2·2
1	2·0	10·0	14·0	2·2

REFLOW SOLDERING				
CASE SIZE	A	B	C	D
1a	3·0	3·0	9·0	2·0
1	3·0	7·0	13·0	2·0

Fig. 5.14 Footprints for flow and reflow soldering of SM electrolytic capacitors.

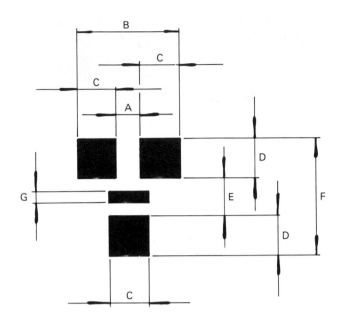

DIMENSIONS:

WAVE SOLDERING							
	A	B	C	D	E	F	G
	0·8	3·4	1·3	1·3	1·2	3·8	0·4

REFLOW SOLDERING							
	A	B	C	D	E	F	G
	1·2	2·6	0·7	1·1	2·6	4·8	0

Fig. 5.15 Footprints for flow and reflow soldering of SOT23 semiconductors.

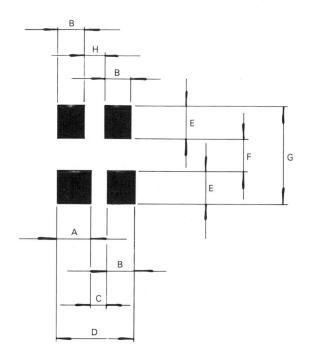

DIMENSIONS:

WAVE SOLDERING								
	A	B	C	D	E	F	G	H
	1·3	1·0	0·6	2·9	1·2	1·2	3·6	0·8

REFLOW SOLDERING								
	A	B	C	D	E	F	G	H
	1·1	0·7	0·8	2·6	0·9	1·1	2·9	1·2

Fig. 5.16   Footprints for flow and reflow soldering of SOT143 semiconductors.

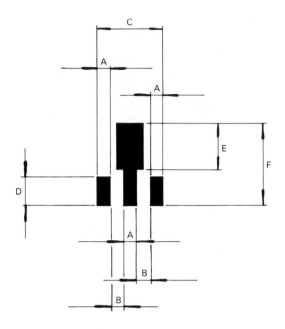

DIMENSIONS:

WAVE SOLDERING							
	A	B	C	D	E	F	G
	1·0	0·5	4·0	1·5	3·5	5·5	2·0

REFLOW SOLDERING							
	A	B	C	D	E	F	G
	0·8	0·7	3·8	1·2	2·6	4·6	2·0

Fig. 5.17 Footprints for flow and reflow soldering of SOT89 semiconductors.

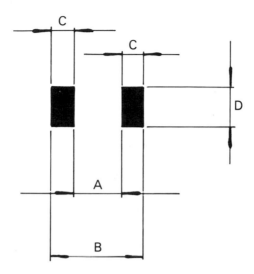

DIMENSIONS:

WAVE SOLDERING				
	A	B	C	D
	2·5	5·0	1·25	2·0

REFLOW SOLDERING				
	A	B	C	D
	2·4	5·2	1·4	1·4

Fig. 5.18   Footprints for flow and reflow soldering of SOD80 diodes.

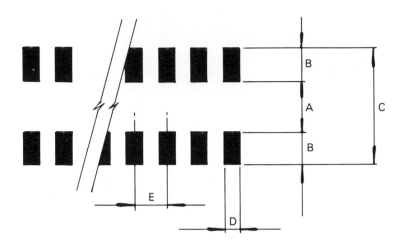

DIMENSIONS:

WAVE SOLDERING					
PACKAGE SIZES	A	B	C	D	E
SO 8 to 16	4·0	1·5	7·0	0·6	1·27
SO 16L to 28	7·8	1·8	11·4	0·6	1·27
VSO 40 to 56	8·0	2·7	13·4	0·5	0·76

REFLOW SOLDERING					
PACKAGE SIZES	A	B	C	D	E
SO 8 to 16	3·8	1·9	7·6	0·9	1·27
SO 16L to 28	7·8	2·0	11·8	0·9	1·27
VSO 40 to 56	8·0	3·1	14·2	0·5	0·76

Fig. 5.19   Footprints for flow and reflow soldering of the range of SO packaged integrated circuits.

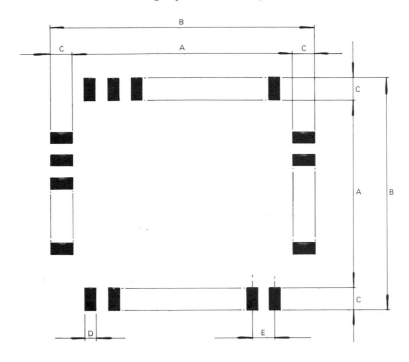

DIMENSIONS:

REFLOW SOLDERING (ONLY)					
	A	B	C	D	E
PLCC 44	14·1	18·7	2·3	0·6	1·27
PLCC 68	21·8	26·4	2·3	0·6	1·27
PLCC 84	26·9	31·5	2·3	0·6	1·27
QUAD 44	10·0	14·0	2·0	0·5	0·8

Fig. 5.20 Footprints for reflow soldering of a range of PLCCs and quad packaged integrated circuits.

Chapter 6

COMPONENT PLACEMENT

K. R. STONE

Formerly of Mullard Mitcham, Mitcham, Surrey, England

6.1 INTRODUCTION

One of the main aims of surface-mounting technology is to put a large number of components into a small space reliably and quickly, i.e., economically.

In more densely populated assemblies, the surface mounted devices (SMDs) can be packed in so tightly that they are almost touching one another. The components are generally small, difficult to handle, and unmarked (although some manufacturers are now introducing marked components). An assembly may contain many different component values and types. Components must therefore be placed accurately and reliably, and if any economic advantage is to be gained—quickly.

A strict discipline must be maintained to avoid mixing components up and placing them in the wrong place. Automated assembly methods are necessary to ensure repeatability, accuracy and economic advantage.

Hand assembly methods are possible but prone to human error and are therefore not generally recommended for use in a production environment.

This chapter initially deals with the principles of hand and machine placement and covers the basic functions that a placement machine must have. The second part of the chapter is a list of currently available machines with a brief specification given for each.

6.2 HAND ASSEMBLY

Hand assembly can be a time-consuming operation requiring a lot of organisation and concentration. The components must be sorted out into labelled containers that allow easy access. Because a large number of very similar looking components are being used, it is important that these do not become mixed up; therefore some means of ensuring that components cannot be accidentally dropped into the wrong container should be devised. There are several methods of achieving this: containers with lids that are only taken off when a component is removed is one method; a tray divided into compartments with a small, known number of components in each section is another method. Any compartment that has the wrong number of components can be checked or discarded. Alternatively, one of the commercially available semi-automatic hand-assembly aids can be used. These generally present one component bin at a time to the operator to avoid errors.

There are two main methods of picking up the components—tweezers and vacuum pipettes. Tweezers are cheap, easy to use and generally readily available. Stainless steel fine point tweezers are recommended for chip resistors, capacitors and discrete semiconductors. Vacuum pipettes are also commonly used; these have pencil-shaped handles with a switch to control the vacuum and a plastic covered pick-up tip that can be rotated to orientate the components. The main advantage of vacuum pipettes is the ease of rotation of the components. The main requirement of this type of assembly method is a steady hand, particularly when placing some of the larger ICs with lead pitches of 25 thou or less on all four sides. For these types of packages and other surface mounted ICs, jigs can be made with cut-outs to locate the components quickly and accurately without too much difficulty.

Hand assembly methods are ideal for assemblies with only a few SMDs which would not justify the use of automatic placement equipment owing to the large proportion of time that would be spent loading and unloading the machine. This usually applies to assemblies that contain only a few surface-mounted ICs. As ICs are normally labelled, there is not the same danger of mixing them up, but there is still a danger of placing them with a 180° mis-orientation. It is generally not a good idea to use hand-assembly methods as a matter of course for prototyping. The time required to hand assemble any but the simplest surface-mounted assemblies is similar to that required to programme a software controlled placement machine. Once the programme has been written, it is a simple matter to run off a few identical prototypes, and repeat orders are easily fulfilled. If the assembly is to go into production, the programming time must be spent in any case, and, if there are layout changes, these can easily be edited into the programme. If prototypes are hand-assembled, there is no guarantee that all have been assembled identically and correctly. When auto-assembly is used, a 'run-off' of the components can be made onto double-sided sticky tape that can then be checked. This ensures that the programme is correct and, by using auto assembly, all the samples will be identical.

6.3 MACHINE ASSEMBLY

For the most economic use of surface-mounting technology, machine assembly methods are recommended. There are various types of machines available now and many new ones being introduced; these range from sophisticated 'hand-assembly' aids, through to hardware controlled multi-headed giants.

Most of the placement machines available have been developed from those used in the thin and thick film hybrid industry for applications involving a few component types on fairly small substrates, 4×4 in. or less. These early machines typically have a moving head picking and placing components onto a stationary substrate. The bulk components are fed to the pick-up position by vibratory feeders. This type of machine typically runs at about 1-2 thousand components per hour. With the introduction of surface-mounting technology on PCBs, three basic machine requirements have changed. First the substrate sizes have increased up to a maximum of about 18 in. square; secondly the number and range of components have increased, and finally the production volumes have increased. New component feed systems have been developed including taping and reeling to cope with the large numbers of components. New machines have evolved that can cater for larger substrates, for more and different component feeders, and that have automatic load/unload facilities to cope with the large throughput typical of a consumer electronic assemblies activity. This type of machine runs at anything from two thousand components per hour upwards.

At the end of this chapter there is a brief summary of some popular machines available. No attempt has been made to recommend a particular machine because each application is different and no one machine is ideal for all situations.

There are, however, several points that should be considered, whichever machine is chosen. Of greatest importance is that a discipline must be imposed to ensure that the right components are in the correct feeder positions. If an incorrect component feeder is loaded on the machine, a large number of rejects can be made before the fault is detected, particularly as surface-mounting is usually one of the first processes to be used, and testing the last. There is an increasing trend towards supplying SMDs on reels of tape, each reel holding from 3000 to 10,000+ components, which can cause difficulties when planning the purchasing for either prototypes or small production batches. There are also some machines that require more than one component feeder position for a given value of component placed on the substrate. For instance, if a 10 ohm resistor is to be placed at opposite ends of a large PCB, the machine may require two 10 ohm reels of resistors loaded at opposite ends of the machine.

Other points to consider which are covered in other chapters are the design limitations imposed by the machine; its accuracy, and situations as mentioned earlier, where the number of component feeders may be reduced by careful design; whether to use adhesive or solder paste; which type of glue, U-V cured or thermally cured; one part or two part and what cure cycle.

6.3.1   Machine Functions

The basic functions that a placement machine has are dealt with in the following seven sections.

6.3.1.1   SUBSTRATE POSITIONING AND FEED

The machine must have some means of positioning the substrate under the placement head(s) so that the SMDs can be placed in the correct position.

6.3.1.2   ADHESIVE OR SOLDER PASTE APPLICATION

Chapter 7 will describe component attachment in detail, but basically there are three main methods of holding the components in position:

(i) Non-conductive epoxy to hold the component in place temporarily until the electrical connections are made by wave or jet soldering. A small amount of adhesive is placed under the centre of the SMDs so that it does not contaminate the terminations and solder pads. If the adhesive is cured by ultra-violet radiation it must be visible to that radiation and applied accordingly.

(ii) If a reflow soldering process is to be used, the components are initially placed and held in the 'wet' solder paste prior to reflow and the forming of the permanent electrical joint. The solder paste is applied to the conductors on the substrate where the terminations of the components are to rest.

(iii) Electrically conductive adhesives can be applied to the component pads on the substrate to provide both the permanent mechanical and electrical joint. This is probably the least widely used method.

The adhesive or solder paste application is usually carried out by one of the following methods:

(i) Printing

Used mainly for U-V cured adhesives, thermally cured adhesives, and solder paste. All the dots of adhesive or solder paste are placed on the entire substrate(s) at once by screen printing or by using a stencil prior to the substrate being put on the placement machine.

(ii) Pressure Dispense

This is a popular method used for thermally cured adhesive and conductive adhesives. Dots of adhesive are dispensed onto the substrate sequentially by pressure being applied to a glue reservoir. This can either be done in a separate glueing cycle or simultaneously with the placement head(s).

(iii) Pin Transfer/Transfer Printing

Primarily used for applying thermally curing adhesive but can also be used for applying solder paste. The adhesive or paste is picked up by dipping a pin or transfer tool into a bath of adhesive or paste—this is then transferred to the placement position. This can either be performed sequentially or, more commonly, when using adhesive, a 'bed of nails' is used to transfer glue to all the required positions simultaneously.

(iv) Chip Transfer

Used only for non-conductive adhesive application. The glue is applied directly to the underside of the chip component by a pin or knife blade immediately before placement; thereby the chip transfers its own adhesive.

6.3.1.3 PARTS FEED

A means of feeding the various components to the pick-up position is required. There are three main methods of component feeding:

(i) Tape

The components are packaged in paper or plastic tapes on reels of typically 5000, as described in Chapters 3 and 4. The paper tape consists of a base tape with slots, in which the components sit, and cover tapes (top and bottom) made of plastic or paper holding the components in place. The plastic tape consists of embossed pockets for the components to sit in and a plastic cover tape to hold them in. There is also aluminium tape which is similar in construction to the plastic tape but the base material is an alu-minised plastic, which has better handling properties (see Chapters 3 and 4). The tapes have indexing holes which are advanced by the tape feeder each time a chip is picked up. There is also a mechanism for either pulling back the top cover tape or cutting the tape to release the component before pick-up.

(ii) Vibratory Bowl

Loose components are placed in vibratory bowl feeders which sort out the components into a single line all in the same orientation. Usually there is one bowl feeder for each pick-up point. A disadvantage of this method is that there is a danger of putting incorrect components in the bowl by mistake. One chip looks much like another, so the contents of the bowl may have to be scrapped. It is also time consuming changing the components in a bowl feeder to another type.

(iii) Pre-loaded Cartridges and Sticks

Components are fed from the sticks or cartridges, usually by vibrators, to the pick-up position. Disadvantages of this method are the limited number of components in the feeder compared with bulk fed or taped and reel. An advantage is that only one bowl feeder is required to fill all the sticks with chip components. Vibratory stick feeders are probably the best way of handling those components that are not available in tape, and other components that are easier to obtain loose, particularly in low numbers, from distributors.

6.3.1.4   PICK-UP HEADS

A head or heads are required to pick up, orientate and centre the components. The head can either pick and place at fixed positions or it can be controlled to pick from a variety of places, and place anywhere over the placement area depending on the machine design.

Most machines use a vacuum pipette to pick the components from the feeder. Centring of the component on the pick-up tip is often necessary to eliminate any positional error of the components in the feeder, particularly in the case of taped components where the component is free to move around in the tape pocket.

The centring is usually carried out by four tweezer arms which either move in simultaneously or in pairs to centralise the component on the vacuum pipette. Usually the vacuum is sensed to check that a component has actually been picked up and also that it has actually been placed at the right time. Many placement machines have pick-up heads that can use a variety of pick-up tools to cater for different size and shape components. The pick-up tool is usually changed under programme control when, for example, changing from chip resistors to SO packs or MELFs.

6.3.1.5   PROGRAMMING

The machine must be able to be programmed with the information on which component to place where. It is desirable to be able to edit and change programmes.

Programming can either be hardware- or software-controlled.

(i) Hardware

Hardware control is used on the high volume production machines where a large number of relatively few different types of assembly are required.

The pick-up heads are moved into positions against stops on a hardware programme plate. On some machines the CAD software used to lay out the PCB can be used to generate numerically-controlled drilling data for the programme plate.

This type of programming is obviously expensive to edit and can take a long time to change.

(ii) Software

Most of the small to medium volume machines use software-controlled programming systems. There are two basic types, on-line and off-line.

On-line systems are more commonly found on the smaller machines. These use a self-teach process where the machine is stepped through the placement cycle under manual control. The machine 'learns' the pick-up and placement positions. The main disadvantage of this method is that the machine cannot be used for production while programming is being carried out.

Most of the medium size machines therefore have the facility of off-line programming. This is usually performed by one of three methods: keyboard entry, digitising, and by direct link to a PCB layout CAD system. All three methods basically feed the placement position co-ordinates into the assembly machine. The pick-up positions must either be fed in separately or the machine will decide for itself the optimum feeder positions.

Generally the programming systems have optimisation routines to choose the best feeder positions and shortest routes for placement.

This type of programming is easy to edit and change and new programmes can be loaded in just a few minutes.

6.3.1.6 COMPONENT SAFETY FEATURES

The machine control must include some means of determining whether a component has been picked up or not, and if it has been placed or not. This avoids missing components. Most machines sense a vacuum on the pick-up tip which indicates whether a component is present. If a feeder is empty, usually the machine re-tries a number of times before signalling a fault. It is also important to sense that the component has been placed at the right time, otherwise components may fall off the pick-up tip in transit and land in the wrong place; meanwhile the pick-up tip may be dipped in the adhesive where the chip should have been. Some machines also have component testing prior to placement which indicates whether the component is the right resistance or capacitance, or the correct polarity if it is a semiconductor. This may be an advantage if a large number of bulk fed components are used. If taped and reeled components are used, they are normally pre-tested by the manufacturer and labelled with the value so that there should be no need for component verification. It is a good idea, however, to test the first few components of a reel at a Goods Inwards inspection, as it is not unheard of for components to be wrongly labelled.

6.3.1.7 CURING SYSTEM

This is not strictly part of the basic machine but is necessary when using the stick and wave solder process to prevent movement of the chip components during subsequent board handling. The type of oven required will depend on the chosen adhesive system; an oven fed directly from the placement machine would be ideal.

If a solder paste reflow method is being used, a reflow system (including a drying oven) will be required instead of a glue curing oven.

6.3.2   Machine Types

The popular machines can be classified into four main groups, as dealt with in the following sections.

6.3.2.1   SEMI-AUTOMATIC ASSEMBLY AIDS

These generally consist of bins containing the different components which are presented in turn for hand pick-up of the components using vacuum tweezers, and the component orientation and placement position is indicated on the board, for hand placement.

6.3.2.2   SEQUENTIAL PICK AND PLACE MACHINES

These usually have a single head which can place a large number of different components at different points on a substrate under the control of a small computer system. There are two main ways of programming this type of machine, off-line and on-line as previously described. There is a further split in the categorisation of the machine types, and this is the distinction between low to medium and medium to high volume sequential pick and place machines. The former are more geared towards the hybrid industry and usually have a single placement head with placement rates of 1-4 K per hour, and handle fairly small substrates. The latter group are faster at 4-16 K per hour and generally have more than one head but only pick up one component and place one component at a time, as opposed to the true multi-head machines which pick up components simultaneously and place them either sequentially or simultaneously. This type of machine has been specially designed for the PCB industry and therefore caters for large substrates.

6.3.2.3   SOFTWARE PROGRAMMED MULTIPLE HEAD MACHINES

These are similar to the previous category but they pick up a large number of components simultaneously and place them under programme control either sequentially or simultaneously. The programming systems are usually off line owing to their somewhat more complex assembly method. These sorts of machines are generally for medium to high production runs as they combine speed and flexibility.

6.3.2.4   HARDWARE CONTROLLED MACHINES

This type of machine is the least flexible but usually the fastest. They generally have multiple heads and pick up a large number of components simultaneously and place them in a single cycle. Although the machines are very fast, they are difficult and expensive to re-programme and are therefore suitable for large runs of the same board assembly.

6.3.3. Survey of Some Popular Machines

6.3.3.1 SEMI-AUTOMATIC HAND ASSEMBLY/TRAINING AIDS

Mamiya Denshi Co. Ltd Superhand Trainer ECM83 (Ref. 1)

This machine (see Figure 6.1) is aimed at the low-volume end of the market, for pre-production runs and training of engineers and/or operators.

Fig. 6.1   The Mamiya Denshi Superhand Trainer ECM83. (Courtesy of Hedinair Ltd, Essex, UK)

Basically, the machine consists of an X-Y plotter controlled by a small computer. The chips are picked up by a vacuum pipette and stepper motors move the head across the placement area. There are bulk component feeders for 12 lanes of chips as standard, and up to 8 magazines can be added for SOT23 semiconductors. Adhesive application can be carried out by the machine, prior to placement, with a dispensing head.

Specifications

Dimensions (Main Frame):	567 mm(L) × 605 mm(W) × 180 mm(H)
Weight:	20 kg
Placement Area:	X axis 360 mm; Y axis 270 mm
Mount Speed:	250 mm/s
Repetition Accuracy:	±0·2 mm (average)
'Z' Axis Travel:	5 mm
Pick-up Sensor:	Yes
Adhesive Application:	Optional head available
Power Source:	AC 100 V + 10%, 50/60Hz
Air Pressure:	5·5 kg (dry air)
Component Types:	Both square edge chip and mini-mold transistor can be handled. No Z axis rotational movement.
Component Feeders	Both rail and hopper feeders for bulk components (12 lanes)

Schlup Chiplacer (Ref. 2)

PCBs are mounted on a microprocessor-controlled X-Y table; adhesive or solder paste is automatically dispensed under computer control; the board is then moved to the placement window where the chip components are manually placed. The whole operation is monitored by a video system. Up to 90 different components can be picked from random access bins; there is also provision for 8 mm tapes (see Figure 6.2).

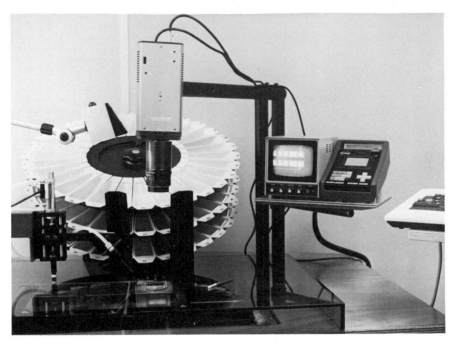

Fig. 6.2 The Schlup Chiplacer. (Courtesy of Blundell Production Equipment Ltd, Coventry, UK)

Specifications

Dimensons:	1700 mm(L) × 1400 mm(W) × 1100 mm(D)
Weight:	125 kg net (excluding keyboard/screen)
Placement Area:	330 mm × 250 mm
Speed:	3600 elements/hour, 800–1000 typical. Operator dependent.
Accuracy:	0·05 mm
Repetition:	0·05 mm
Adhesive Application:	Programmable adhesive dispenser
Power Unit:	220 V, 50 Hz, 550 VA
	110 V, 60 Hz, 550 VA
Air:	6 bar
Component Types:	All surface mounted components and chip carriers.
Component Feeders:	Total 90 (3 levels of 30 each) random access bins
Programming:	On-line or off-line

Universal Logplace (Ref. 3)

The Logplace is based on the logpoint machines used for the hand assembly of leaded components. Under microprocessor control the correct part is located from tape or one of 63 pater-noster trays and the placement position is indicated to the operator.

The machine is programmed on line using a joystick and keyboard. A vacuum pencil is used to handle the components.

Model 4511A   **Specifications**

Dimensions:	2438 mm × 1067 mm × 2033 mm
Weight:	175 kg
Maximum PCB Size:	254 mm × 254 mm
(with rotation capability)	
Placement Speed:	Operator dependent, up to 1000 components per hour.
Adhesive Application:	Off-line
Power Input:	110-120 V, 60 Hz, 15 A STD
	220-240 V, 50 or 60 Hz Optional
Air:	75 psi (min)
Component Feeders:	63 bin pater-noster magazine standard up to 15, 8 mm tape feeders. SOT-23, SO pack stick feeders optional
Programming:	On-line

6.3.3.2   SEQUENTIAL PICK AND PLACE MACHINES (LOW–MEDIUM VOLUME)

Dyna/Pert–Precima Ltd MPS 100 (Ref. 4)

There are two main machines based on the MPS 100—one with a tweezer head and the other with a turret head.

The turret head system is a microcomputer controlled pick and place machine for hybrids and small PCBs. The maximum substrate size is 4 × 4 in. but up to 6 × 6 in. can be accommodated if some feeders are eliminated. It has a single head with four or eight pick and place tools. The vacuum head picks and places up to 20 different values of components at a cycle rate of 3600 per hour.

The tweezer head version is similar, but it has a single head with a tweezer mechanism for centring components, and it is slower with a cycle rate of 2,500 components per hour.

Both machines can handle components in tape, vibratory feeders, and various die feeder systems.

Adhesive application can be accomplished by transfer printing, using a special tool, but at the expense of placement rate.

Programming is by a self-teach system and the completed programmes are stored on microcassette.

Specifications

Dimensions:	1·65 m(H) × 1·22 m(W) × 0·91 m(D)
Weight:	136 kg
Placement Area:	Normally 102 mm × 102 mm
	Up to 102 mm × 152 mm if some feeders are eliminated.
Speed:	Cycle rate 2,500 (tweezer head version), 3,600 (turret head version) components per hour.

Accuracy:	±0·003 in.
Adhesive Application:	By transfer print tool mounted on the placement head.
Pick-up Sensor:	Vacuum sensing of component presence before and after pick-up. Automatic re-try on up to three additional components.
Power:	115-240 V
Air:	60 psi
Component Types:	Chip resistors, capacitors, diodes, SOT23, SO packs, leaded/leadless chip carriers.
Component Feeders:	8 and 12 mm tapes, flat vibratory feeders, ski-slope feeders, and various die feeder systems. Up to 20 different component types.
Programming:	Self-teach.

Dyna/Pert—Precima MPS 118 (Ref. 5)

The MPS 118 was developed as a large PCB machine capable of placing a wide range of SMDs onto boards up to 18 in. square.

The placement head uses a vacuum nozzle and centring tweezers to pick components from tape or vibrator and place on the Y-table. Up to 60 component feeders can be loaded on the machine. Glueing can be carried out with a dispense head but this occurs before each placement which reduces the production speed. Programming is carried out on-line with a self-teach hand-held keyboard.

Specifications

Dimensions:	2·56 m(L) × 1·25 m(H) × 1·6 m(W)
Weight:	400 kg
Placement Area:	457 mm × 457 mm
Placement Speed:	2,000 components per hour cycle rate
Accuracy:	±0·2 mm
Adhesive/Solder Paste Application:	Pressure dispensing by independent head depositing before each placement with attendant reduction in production speed.
Pick-up Sensor:	Vacuum sense of component presence. Missing component will result in 3 automatic re-tries.
Power:	100–240 V; 2 KA; 47–63 Hz
Air:	552–828 kPa (80–120 psi)
Component Types:	SOT components, SO packs, chip capacitors, resistors and diodes, leaded and leadless chip carriers.
Component Feeders:	60 tape reels (8 mm) or 30 tape reels (12 mm) or 30 vibratory channels.
Programming:	On-line self-teach.

Excellon MC30 (Ref. 6) (Figure 6.3)

The main selling point of this machine is its flexibility of feeder type. It can handle tapes, bulk, waffle packs, sticks and cartridges. Up to 100 different component types can be fed in stick and vibrator, but a lot less in tape. The machine is programmed on a step-through self-teach system and programmes are stored on floppy disc. The head has a simple vacuum pick-up tip which moves along the X axis and the substrate is positioned in the Y direction on a motor driven table. A tweezering mechanism will shortly be available for ensuring components are accurately positioned, particularly when picked from tape.

There is no separate glueing head, although a tool can be fitted to the pick-up head which can be dipped in epoxy which is then transferred to the substrate. This reduces the typical 2000 per hour placement rate.

Fig. 6.3   The Excellon MC30. (Courtesy of Excellon International, West Sussex, UK)

Specifications

Dimensions:	52 in.(H) × 66 in.(W) × 29 in.(D)
Weight:	720 lb
Maximum Board Size:	457 mm × 610 mm
Placement Speed:	2,000 parts/hour typical
Accuracy:	±0·05 mm
Repeatability:	±0·05 mm
Power Input:	100−240 V

Air: 30 psi (if die centring device is required)
Component Types: Size range 1 mm^2 to 32 mm^2
Programming: Self-teach, programmes stored on floppy disc.

Panasert (Ref. 7)

Several robot type pick and place machines are included in the Panasert range for handling large SMDs and surface mounted ICs in particular. The machines have a pick and place head that takes components from tape feeders and places them anywhere on the board. Glueing can be done using a dispense head, but this operates alternately to the placement head, thus reducing the speed.

Specifications

Model	NM 2521	NM 2520	NM 2503
Dimensions:	670(W) × 1,440(D) × 1,674(H)	1,100(W) × 2,220(D) × 1,220(H)	1,650(W) × 1,450(D) × 1,400(H)
Weight:	Approx. 500 kg	Approx. 300 kg	Approx. 1,200 kg
PC Board Size:	Max. 200 × 200 250 × 330 special Min. 90 × 60	Max. 330 × 250 Min. 90 × 60	Max. 330 × 250 Min. 90 × 60
Placement Time:	Approx. 1·9 s Dispensing & placement approx. 2·8 s Placement & Placement approx. 1·6 s	Approx. 4·1 s (2 adhesive dispensing points/time)	Approx. 1·0 s
Component Feeders:	Max. 32 8 mm taped Max. 28 12 mm taped Max. 16 32 mm taped	Max. 10 12 mm taped Max. 8 32 mm taped	20 flat pack ICs
Controller:	Panadac 771	Panadac 771	Panadac 781

(Dimensions in mm unless otherwise stated.)

Universal Omniplace (Ref. 8)

The model 4621A placement system uses two overhead X-Y positioning heads, operating independently to place SMDs on PCBs or ceramic substrates. The machine can be a stand-alone unit or part of an in-line system.

The basic unit has 40 component feeders and up to four more feeder units can be added giving a total of 192 feeder positions. The variety of component feeders includes bulk, tape, sticks and matrix trays. Component verification is available as an option.

Specifications

Dimensions: 66(W) × 54(D) × 70 in.(H) (167 × 137 × 177 cm)
Board Size PCB: min. 5·9 × 4·25 in. (150 × 108 mm)
                         max. 18 × 16 in. (457 × 406 mm)
Ceramic Substrate: min. 2 × 2 in. (51 × 51 mm)
                         max. 8 × 8 in. (203 × 203 mm)
Placement Cycle Rate: 4500 chip components per hour when placing on 4 × 4 in. (102 × 102 mm) circuit with an average placement head movement of 8 in. (203 mm).

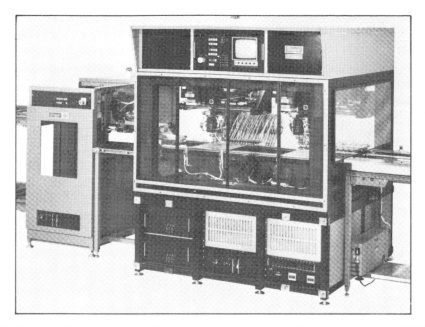

Fig. 6.4 The Universal Model 4621A Omniplace. (Courtesy of Universal, New York, USA)

Z Axis Travel:	1·2 in. (30 mm)
Component Sensor:	Verifies component pick-up and placement. Detects a missing component then re-tries a programmable number of times to pick up a component.
Component Size Range:	0·030 in.sq.–1·25 in.sq. (0·76–32 mm) depending on type of tooling selected.
Component Input Method:	Bulk, tapes, sticks, matrix trays.

Zevatech Automatic Placement and Assembly Systems for Micro-components (Ref. 9)

Zevatech offer a range of machines from one assembly station with production capacities of 2400 components per hour to comprehensive 'tailor-made' production lines with several assembly stations capable of more than 10,000 components per hour.

The modular design systems use pick and place heads to assemble a wide variety of components onto substrates up to 20·5 × 20 in. Up to 128 component feeders per assembly head can be used. These include tape, vibratory stick and bowl feeders. Placement can be preceded by a glue dispensing station.

Specifications

Refer to manufacturer.

6.3.3.3   SEQUENTIAL PICK AND PLACE MACHINE (MEDIUM-HIGH VOLUME)

Amistar SM-1000 (Ref. 10)

The Amistar is a relatively high speed machine placing up to 14,400 components per hour using a system of six rotating heads. The machine can handle 64 8 mm tapes with provision for expansion to 128 tapes. Both manual and automatic substrate feed versions are available.

Specifications

Dimensions:	195·6 (W) × 139·7 (D) × 160·0 (H) cm
Weight:	454 kg
PC Board Size:	Manual 16 in. × 16 in.
	Auto Max 15 in. × 15 in.
	Min 3 in. × 3 in.
Placement Speed:	14,400 placements per hour
Placement Accuracy:	0·004 in.
Power:	115 VAC, 60 Hz/230 VAC, 50 Hz, 1·8 kVA
Air:	Un-oiled dry air min 80 psi at 12 CFM max.
Component Types:	Chip capacitors, resistors and SOT23s
Component Feeders:	Sixty-four 7 or 11 inch, 8 mm tape reels.
Programming:	Programmes stored on floppy disc with Trackball Programmer.

Dyna/Pert-Precima Ltd MPS-500 (Refs. 4 and 11)

This is a larger machine capable of placing up to 120 different component types at a cycle rate of 6,000 per hour, over an area of 18 × 14 in.

Programming is off-line, using floppy discs as the storage medium. There are two servo-driven component feeder carriages that independently present components to the pick and place heads. There are two vacuum heads with centring tweezers; while one is picking up a component, the other is placing a component onto the X-Y table.

Epoxy can be applied using a pressure dispense head simultaneously with placement cycle at a prior station and requires no additional time. The machine is shown in Figure 6.5.

Specifications

Dimensions:	2·9 m(L) × 1·57 m(H) × 2·11 m(W)
Weight:	1100 kg
Placement Area:	457 mm × 356 mm
Placement Speed:	6,000 components per hour test rate
Accuracy:	±0·2 mm
Adhesive Application:	Pressure dispense by independent head on previous board position.
Pick-up Sensor:	Vacuum sense of component presence. Automatic re-try for up to three components.
Power:	100–240 V; 3·84 kVA; 47–63 Hz
Air:	552–828 kPa (80–120 psi)
Component Types:	SOT components, SO packs, chip capacitors, resistors and diodes, leaded and leadless chip carriers.

Component Feeders: 120 tape reels (8 mm) or mix of 8, 12 mm tape feeders and SO pack stick feeders.

Programming: Off-line programming unit with RS232 interface which downloads to the MPS500.

Fig. 6.5 The Dynapert Precima Microplace System (MPS) Model 500. (Courtesy of Dynapert Precima, Colchester, Essex, England)

Fuji Machine Manufacturing Co. Ltd. CP II (Ref. 12)

This is a new generation machine from Fuji and uses a built-in camera and computer to replace mechanical centring fingers. Using a system of rotating heads, the components are picked up one at a time, the camera senses the rotation and displacement in the X and Y axes on the pick-up tip, the head is then rotated to the corrected orientation and the X-Y table positioned to its corrected position under control of the computer, and then the component is placed on the PCB. This system should result in faster placement, increased accuracy and less maintenance than a similar system using mechanical tweezering.

Specifications

Dimensions: 3240 (L) × 1650 (D) × 1600 (H) mm
Weight: 2800 kg
PC Board Size: 250 × 330 mm max (standard)
360 × 470 mm max (option)
Placement Speed: 14,400 max. per hour
Placement Accuracy: ±0·1 mm
Power Supply: 11 kVA, 3 phase AC, 200/210/230/380/415/460/480 V, 50/60 Hz

Air Supply:	4 kg/cm² (60 psi) min.
Component Types:	Resistors, capacitors, SOT23s, MELFs, tantalums, trimmers, coils and SOICs.
Component Feeders:	Up to 100 types in 8, 12 or 16 mm tape.

Nitto Kogyo STM-8 (Ref. 13)

This machine is designed to place bulk fed MELF components onto PCBs using a single head. The loose components are simply poured into the hoppers and the STM-8 places them at a rate of up to 7200 components per hour.

Specifications

Dimensions:	700 (W) × 850 (D) × 1050 (H) mm
Weight:	300 kg
PC Board Size:	200 × 200 mm max.
Placement Speed:	7200 max. per hour
Power Supply:	AC 200 V, 3 phase, 50/60 Hz, 2 kVA
Air Supply:	5 kg/cm², 100 litres/min.
Component Types:	MELF resistors, capacitors, cylindrical ceramic chips.
Component Feeders:	Hoppers for bulk only, 40 feeders

Panasert (Ref. 7)

There is a large range of Panasert machines in the sequential medium to high-volume pick and place machines. They all work in a similar way with a pick and place head moving in the Y direction and placing onto the substrate that moves along the X direction. The components are held in tape on a carriage that moves in the X direction to position the relevant component under the placement head. The glueing is performed by a dispense system in parallel with the placement cycle. The various machines can have twin heads, carousel component feeder carriages and different options on the types of component that can be handled. They are built as modules so that several options can be joined to give a combined system that will cover a wide range of components. (See p. 155 for Specifications.)

Siemens—Automatic Placement System MS-72 (Refs. 14 and 15)

This machine is a flexible system for placing a wide range of SMDs from a wide range of feeder systems. The system can have either a single placement head, two heads operating in tandem or one placement and one adhesive dispensing head (see Figure 6.6).

The placement head moves in both the X and Y axes and has a vacuum nozzle and tweezering mechanism. The tweezers are electrically insulated to allow component verification.

The component feeder systems available are vibrators, vibratory sticks and tapes. A mixture of these can be used and a maximum of 210 different component types can be loaded.

The programming can either be self-teach or off-line.

Specifications

Dimensions:	1·7 m(L) × 1·1 m(W) × 1·4 m(H)
Weight:	800 kg

Specifications—Panasert

Model:		NM-2501B	NM-2501C	NM-2501D	NM-2511	NM-2502	NM-2534	NM-2540	NM-8260
Dimensions:		3,270(W)× 1,660(D)× 1,220(H)	3,270(W)× 1,660(D)× 1,220(H)	3,270(W)× 1,660(D)× 1,220(H)	1,690(W)× 1,660(D)× 1,220(H)	1,650(W)× 1,600(D)× 1,435(H)	1,350(W)× 1,650(D)× 1,480(W)	1,925(W)× 2,150(D)× 1,180(H)	3,480(W)× 1,030(D)× 1,850(H)
Weight:	Approx.	1,500 kg	1,500 kg	1,500 kg	2,000 kg	1,500 kg	800 kg	1,300 kg	1,800 kg
PC Board Size:	Max.	330×250	330×250	330×250	330×250	330×250	330×250	100×120	330×250
	Min.	90×60	90×60	90×60	90×60	90×60	90×0	20×20	150×80
Placement Time:	Approx.	0·6 s	0·8 s	0·6 s 8 mm taped 0·8 s 12 mm taped	0·3 s (0·6 s)	0·5 s	0·6 s	0·6 s	0·6 s
Component Feeders:		60	48	48	50 (100 in the single head mode)	80	20	20	80 (Cylindrical chip components)

(All dimensions in mm unless otherwise stated.)

Fig. 6.6   A Siemens Automatic Placement System. (Courtesy of Siemens Ltd, Middlesex, UK)

Placement Area:	270 × 240 mm (standard)
	350 × 310 mm (maximum)
Placement Speed:	4,200 uph (50 × 50 mm substrate)
	(8,000 uph with double-head system)
Placement Accuracy:	±0·08 mm
Adhesive Application:	Simultaneously on separate head
Power Source:	220/380 V, 50 Hz, 2 kVA
Air Supply:	6 bar
Component Types:	Chip, MELF, SOT, SO, DIL, etc. with dimensions from
	1·2 × 1·8 × 0·4 mm to 12 × 12 × 6 mm
Programming:	On-line or off-line

Siemens HS-180 (Ref. 16)

Designed for placement volumes in excess of 20 million components per year where a wide range of components will be used. Up to 180 different components on 8 mm tape can be used at once. Other bulk and magazine fed methods can be used as well, including waffle packs. The machine can be equipped with an automatic elevator feed mechanism.

Specifications

Placement Area:	Min. 50 × 50 mm
	Max. 460 × 460 mm
Placement Speed:	9000–1200 components/h
Placement Accuracy:	±0·08 mm
Power Supply:	220/380 V, 50/60 Hz, 6 kVA
Air Supply:	6 bar
Component Types:	Chip resistors, capacitors, MELF and Mini MELF,
	SOT23, 89, 143, SO pack, chip carriers, etc.

TDK Avimount RX1 and FX4 (Ref. 17)

TDK offer two machines with placement rates of 12 K components per hour. The RX1 uses multiple heads on a rotary disc to place 20 different components sequentially. Several of these machines may be linked in-line to increase the number of different components that can be mounted.

The FX4 can mount up to 80 or 120 different components sequentially on to PCBs by sequencing components before placement.

Specifications	**RX1**	**FX4**
Dimensions:	7·2 (L)×1·7 (W) ×1·6 (H) m	4·6 (L)×2·8 (W) ×1·4 (H) m
Weight:	1500 kg	1800 kg
PC Board Size:	320 (L)×250 (W) mm max. 100×70 mm min.	320 (L)×250 (W) mm max. 150×100 mm min.
Placement Speed:	12000 parts/hour (when X,Y movement is 20 mm max.)	12000 parts/hour (when X,Y movement is 15 mm max.)
Placement Accuracy:	±0·2 mm	±0·2 mm
Adhesive Application:	Optional screen printer	Dispenser
Power Supply:	200 Vac, 3 phases, 5 kVA	200 Vac, 3 phases, 5 kVA
Air Pressure:	5 kgf/cm^2	5 kgf/cm^2
Component Types:	Capacitors, resistors, transistors, diodes and inductors	Capacitors, resistors, and SOT23s
Component Feeders:	20, 8 mm tapes	80 or 120 tapes

Universal Onserter (Ref. 18)

The Model 4712A placement system has a turret-style placement head and vacuum pick-up nozzles located at 90° intervals to perform the following operations in rotation:

(i) extract chip from tape;
(ii) verify component;
(iii) eject faulty component;
(iv) orientate and centre the chip;
(v) place the component on the board.

The component feed system consists of a carousel of 64 tape reels.

There is an optional dispense head available to apply epoxy or solder paste which operates without reducing the placement speed.

6.3.3.4 SOFTWARE PROGRAMMED MULTIPLE HEAD MACHINES

Philips MCMI-SM and MCMII (Ref. 19)

These two software controlled machines work on the similar principle of simultaneous pick-up of 32 components per station and sequential placement onto the substrate. The MCMI-SM has one station of 32 heads and 32 tape feeders giving placement speeds of more than 10,000 per hour (see Figure 6.7). The MCMII can have up to three 32 head stations giving placement rates of 33,000 per hour and greater (see Figure 6.8).

Fig. 6.7 A Philips MCM1 Software controlled placement machine. (Courtesy of Mullard Ltd, London, UK)

Fig. 6.8 A Philips MCMII Software controlled placement machine. (Courtesy of Mullard Ltd, London, UK)

Glue is applied to the underside of the chips immediately after pick-up by a row of pins pushing through a bath of epoxy until they make contact with the component. The epoxy on the top of the pins is transferred onto the centre of the underside of the chips. The chips are then placed on the substrate sequentially under software control.

System Type No.	MCMI-SM	MCMII
Max. PC Board Size*:	210 × 320 mm	210 × 320 mm
Placement Speed:	>10,000 typ. per hour	>33,000 typ. per hour
No. of Placement Stations:	1	1 to 3
Cycle Time:	11·5 s typ.	10·5 s typ.
Max. Boards/h:	310	340
Max. Components/Board:	32 × 15 = 480	96 × 15 = 1440
Glue Unit:	Integral	Integral
Component Tapes:	32 × 8 mm	32 × 8 mm or 20 × 12 mm (in development)

*In some cases larger boards may be handled but the figures given represent the maximum placement area usable.

TDK CX1 (Ref. 17)

Up to 20 different taped chip components per head can be mounted onto a PCB in 7·4 seconds. Unit can be linked together to give up to 200 components placed at the same time.

Specifications

Dimensions:	8·1 (L) × 1·4 (W) × 1·4 (H) m
Weight:	1000 kg (main unit and controller)
PC Board Size:	330 (L) × 250 (W) mm max.
	125 × 75 mm min.
Placement Speed:	7·4 secs/20 pieces (per unit)
	(when movement in X,Y is less than 25 mm max.)
Adhesive Application:	Optional screen printer
Power Supply:	200 Vac, 3 phase 5 kVA
Air Supply:	5 kg/cm^2
Component Types:	Wide range including flat, cylindrical and SOT23 components.
Component Feeders:	20 types, 8 or 12 mm tapes
Programming:	Off-line using floppy disc.

6.3.3.5 HARDWARE CONTROLLED MACHINES

Panasert Mm (Ref. 7)

Up to 200 SOT23s and 1206 chip components loaded into magazine sticks can be placed simultaneously in a 5 second cycle onto adhesive placed by the pin transfer method. The system is modular allowing several placement heads to be placed side by side to give a very high volume production system. Board handling, inspection and adhesive curing modules can be added to complete the assembly line.

The magazine sticks hold the components in a vertical stack. A pin pushes up on the components and pushes the top component onto the glued board. The position of the stick determines the position of the component on the board which is therefore restricted to a 2·5 mm grid.

Specifications

Model	NM8270
Dimensions:	5,500–12,000 mm(L) × 1,840 mm(D) × 1,480 mm(H)
Weight:	2,000–7,000 kg (approx.)
PC Board Size:	210 × 210 max., 100 × 80 min. (mm)
Placement Time:	5 s approx.
Input:	Max. 200 chips per placement station
Chip–chip Distance:	2·5 mm grid

Philips MCMI-HM and MCMIII (Ref. 19)

These two machines are hardware controlled and therefore recommended for high volume production runs.

Like the Philips software controlled machines these too are based on the 32 head stations but the difference is that these are controlled by a programme plate with stops which determine the placement position. The component feeder and glueing mechanism is the same but the placement is simultaneous rather than sequential.

The MCMI-HM has one 32 head station giving placement rates of 15,000 per hour whereas the MCMIII can have between 4 and 12 stations which give placement rates of between 180,000 and half-million per hour (see Figure 6.9).

Fig. 6.9 A Philips MCMIII Hardware controlled placement machine. (Courtesy of Mullard Ltd, London UK)

Specifications

	MCMI-HM	MCMIII
Max. PC Board Size*:	210 × 320 mm	210 × 240 mm
Placement Speed:	15,000 per hour	550,000 per hour
No. of Placement Stations:	1	1 to 12 (usually not less than 4)
Cycle Time:	8 s	2·5 s
Max. Boards/h:	470	1440
Max. Components/Board:	32 × 1 = 32	32 × 12 = 384
Glue Unit:	Integral	Die plate module
Component Tapes:	32 × 8 mm	32 × 8 mm or 20 × 12 mm (in development)
Indexing of Boards:	10 mm	80, 160, 240 mm

*In some cases larger boards may be handled but the figures given represent the maximum placement area usable.

Nitto Kogyo RUTI-SERT STM-2 (Ref. 20)

This is a mass placement machine that mounts up to a hundred cylindrical and MELF type components simultaneously for each hopper unit. The components are bulk fed using a hopper fed system. The number of hopper units can be increased to give very high placement rates.

Specifications

Dimensions. Mounting Machine:	3200 (W) × 950 (D) × 2270 (H) mm, Weight 1200 kg
Control Box:	700 (W) × 800 (D) × 2000 (H) mm, Weight 150 kg
PC Board Size:	300 × 350 mm max.
Placement Speed:	up to 45,000 per hour for 1 hopper unit
Power Supply:	100 Vac, 0·5 kW, single phase 200 Vac, 2 kW, 3 phase, 50/60 Hz
Air Supply:	5−6 kg/cm², 200 litres/min.
Component Types:	Melf and cylindrical components
Component Feeders:	Bulk feed hoppers, 100 component types per hopper, optional number of hoppers.

The information provided in the preceding text has been compiled from data sheets from the machine manufacturers and agents.

The survey is by no means definitive and no guarantee can be given as regards accuracy or availability. The rate of change in this area is extremely fast; new machines and modifications are being introduced rapidly. For up to date specifications it is advisable to contact the supplier.

REFERENCES

1 Superhand ECM83, Leaflet FL5408/7 No. 3, Hedinair Ltd (UK).
2 Schlup Chiplacer, Leaflet, G. E. Schlup & Co., Switzerland, December (1983)
3 Universal Logplace, Bulletin 4511A, Universal Instruments Corp, USA, July (1984).
4 Engineering Data Sheets, MPS100-TU, MPS100-TW, MPS100-TC, MPS500, Dyna/Pert Division, USM Corp. USA, January (1984).
5 Machine Specification, Machine Model MPS118, Spec. No. 4493A, Dyna/Pert Division, USM Corp. USA, July (1984).

6 'Feeder Flexibility Aids PCB Assembly', *Electronics Manufacture and Test,* June (1984).
7 Panasert General Catalogue, Dec. No. TI–088–8 0459, Matsushita Electric Trading Co. Ltd, Japan.
8 Universal Omniplace, Bulletin 4621, Universal Instruments Corp., USA, August (1984).
9 Zevatech Automatic Placement and Assembly Systems for Micro-components, Zevatech AG, Switzerland.
10 Amistar SM-1000 Accu-Place Surface Mount, Chip Placement Machine, Amistar Corporation leaflet.
11 Machine Specification, Machine Model MPS500, Spec. No. 500, Dyna/Pert Division, USM Corp, USA, January (1984).
12 Fuji Circuit Board Assembler Series, Bulletin No. 850701, Fuji Machine Mfg. Co. Ltd.
13 Nitto Single Mounting Machine, Model STM-8, Nitto Kogyo Co. Ltd, Tokyo.
14 'Setting the Pace for PCB Placement', *Electronics Manufacture and Test*, July/August (1984)
15 'Automatic Placement System for Surface-mounted Components', Catalogue No. MCH019. 02843, Siemens Ltd, Fed. Republic of Germany.
16 Automatic Placement System HS-180, Data sheet MCH 02311843, Siemens, West Germany.
17 Avimount RX1, FX4, CX1 Data Sheets, 85/1 5M, TDK Corporation, Tokyo.
18 Onserter Profile, Universal Instruments Corp., USA (1984).
19 'Automatic Placement Systems for Surface Mounted Devices', Publication No. 12800701, Mullard Ltd, UK, January (1984).
20 Automatic Placement Mounting Machine, Ruti-sert STM-2, Nitto Kogyo Co. Ltd, Tokyo.

Chapter 7

COMPONENT ATTACHMENT

J. F. PAWLING

Mullard Mitcham, Mitcham, Surrey, England

7.1 INTRODUCTION

Successful soldering of components on a substrate demands that the devices are placed and then held in their correct positions ready for the following soldering process. With conventional components the insertion of tag terminations or wire leads through holes in a printed circuit board provides a convenient method of ensuring that the components are retained in their correct positions. This self-jigging is not possible with surface mounted components and other techniques are necessary. When thick and thin film hybrids first used SMDs, the usual method was to rely on the solder paste previously applied to the substrate surface to hold the components in place until the substrate could be heated on a hot plate to melt the solder. With the desire to place components on both sides of a substrate, and to use non-rigid and poorer thermally conductive substrate materials such as polyester and epoxy fibreglass, other methods need to be considered.

This chapter will discuss some of the many ways in which surface mountable components can be attached to a substrate. These include conductive glues, non-conductive glues, hand, flow and reflow soldering. The post-cleaning of substrates is also included in this chapter.

7.2 OVERVIEW OF ATTACHMENT METHODS

Figure 7.1 shows some of the methods used to attach components to a substrate. With a surface mounted component one has eliminated any direct method of mechanical retention such as crimping over of the tags or leads. The electrical connection may be the only means of mechanical retention, the solder or electrically conducting adhesive being directly applied between the substrate conductor pads and the component's terminals.

Alternatively, the components can be attached to the substrate by an electrically non-conductive glue and the soldering performed as a subsequent operation. Each of the methods will be considered in turn. In addition to describing the method, the advantages and limitations will be indicated wherever possible.

7.3 REFLOW METHODS

The main characteristic of reflow soldering is that the heat is applied separately after the application of the solder. Solder can be applied as a paste or cream, as a

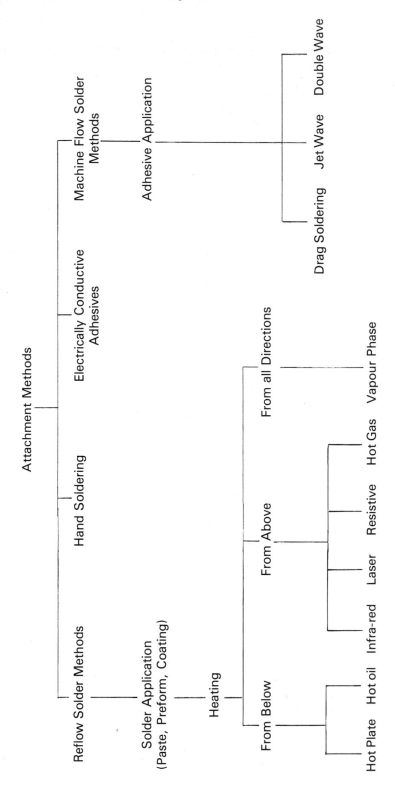

Fig. 7.1 The range of methods used to attach components to a substrate.

solid preform or as a dipped coating. The most popular method of applying solder is to use a solder cream. The following sections describe the selection of such a cream, its application and predrying prior to the application of heat. Various methods of applying heat to re-melt the solder are then described.

7.3.1 Selection of the Solder Cream

Solder creams or pastes consist of a dispersion of powdered solder alloy particles, together with a flux and a solvent. Additional activating agents may also be added. The metal, while accounting for 75 to 90 per cent of the weight, will occupy less than half of the volume. As a result the metal particles are not in direct contact provided the paste has been homogeneously distributed. A whole variety of alloys can be selected offering melting points ranging from 180°C to 300°C + . Most commonly used are Tin/Lead compositions melting in the range 180-210°C, but, where leaching of conductor metals such as palladium silver is a significant problem, solders containing silver, gold and indium can be used. A silver containing alloy may also be beneficial when greater thermal fatigue strength is required. It is not the intention in this book to discuss the selection of solder creams in detail. The reader is referred, particularly, to Chapter 5 of Reference 1 and to Reference 2 which contains an exhaustive bibliography of all the published work on soldering. It is necessary that the user chooses a cream which contains solder powders with the correct particle size and shape. A powder with particles which are too large will cause problems if trying to screen print the paste onto the substrate. To ensure that the paste will pass easily through the screen, the opening should be approximately three times the powder particle size. Too fine a powder, on the other hand, increases the surface area and the chances of oxidation, thus preventing the solder flowing together and increasing the likelihood of solder balling. The tendency to form solder balls is also a function of the solvent used in the paste.

7.3.2 Application of the Solder Cream

Solder cream is applied to the conductor pads of the substrate prior to placing the component, or to the underside of the component itself. The solder cream can be applied by a variety of methods including individual syringe application to the substrate conductor pads, or selected area coverage of a substrate using screen printing and stencilling, or very specific placement employing pin transfer methods. For large scale production the most common technique is screen printing, but pin transfer and dispensing will be briefly considered first.

7.3.2.1 PIN TRANSFER

Although pin transfer techniques can, in principle, be upwards, i.e., against gravity, or downwards, the latter is more popular. The cream or paste is transferred from the main reservoir stock by dipping the pin into it, removing the pin and then transferring the paste to the product. The amount of paste is determined by the shape and size of pin, the properties of the cream—particularly viscosity—how far, and at what speed the pin is dipped into the reservoir and the exact method of transferring the paste from the pin to the product. Typically, up to half a milligram of paste can be transferred. An advantage of pin transfer methods is that the substrate does not have to be flat, as is the case with most other methods.

7.3.2.2 PRESSURE DISPENSING

Solder cream can be dispensed manually or pneumatically and, if required, automatically through hypodermic needles held at an angle of about 30-60° to the substrate. The viscosity of the cream needs to be higher than for pin transfer (approximately 200 Pa.s rather than 100) and care is required in choosing the correct diameter of needle and in controlling the dispensing pressure used if consistent results are to be obtained. A possible problem is that the needles can become clogged unless the lead-in shape to the needle entrance from its paste reservoir is smooth and funnelled and that the paste particles are spherical or elliptical rather than needle-shaped. The advantages of dispensing are that it is flexible so that new layout designs can be handled immediately and also that a variable amount of cream can be applied. For integrated circuits, recommended amounts of solder paste per joint using this method[3] are:

SO (small)	0·5—0·75 mg
SO (large)	0·75 mg
VSO-40	0·75—1·0 mg

7.3.2.3 SCREEN PRINTING

Screen printing is applicable to flat substrates, and is the most popular method of applying solder creams, particularly for the more complex circuits involving close component spacings and a large variety of components for attachment. The usual method is to use an off-contact screen printer in which the downward pressure of the squeegee deflects the screen onto the substrate. As the squeegee passes over the screen, paste is pressed through the mesh openings and onto the substrate. The snap-off of the screen, plus the surface tension of the paste, ensures that it is left on the surface of the substrate once the squeegee has passed. Figure 7.2 illustrates this. Only parts of the substrate require solder paste. To

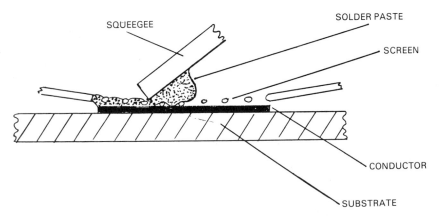

Fig. 7.2   The application of solder paste to the substrate using an off-contact screen printer.

provide the required deposition pattern the screen mesh supports an emulsion barrier which only allows paste to pass through holes previously introduced. The pattern of holes is obtained by exposing the emulsion to ultra-violet light through a photographic film of the required deposition pattern, the emulsion being cured where exposed to U-V and left uncured elsewhere. The uncured emulsion is washed away leaving the required pattern. The screen mesh is generally constructed of stainless steel or polyester, and both the filament diameter and the size of the opening can be varied. The number of apertures per inch or cm is described as the Mesh Number or Count. This can typically vary from 40 to 165 per inch. For application to printed circuit boards, a mesh count of 80 would be suitable[3], while, for ceramic substrates, a finer screen could be used. The wet layer thickness of the screened cream depends on the overall thickness of the gauze and the emulsion, and would be typically 150-250 μm for printed circuit boards. A thickness of around 150 μm is obtained with an emulsion backed 80 mesh steel gauze (30 μm emulsion). The dry later thickness will be similar, as the quantity of solvent used is low.

7.3.3 Component Placement and Predrying

The solvents included in the make-up of the solder paste will gradually evaporate and reduce its tackiness. Components should therefore be assembled as soon as possible after applying the paste to the substrate. Typically, this means within the same day. Once the components have been mounted, it is desirable to reflow solder as soon as possible to prevent oxidation of the metal powder particles and the formation of solder balls, although some manufacturers claim that their pastes can be left for several days. The presence of solvent during soldering is likely to cause solder spattering, the formation of solder balls, and shifting of the position of the placed components. To avoid these problems it is desirable to predry the solder paste after component placement to evaporate the majority of the solvent. The time and temperature are interdependent and also depend on formulation of the solder cream, but it could typically be half an hour at 80-90°C or 10 minutes at 120°C.

7.3.4 Reflow Heating Methods

As stated already, the characteristic of reflow soldering is to apply heat separately after the application of the solder. Heat can be supplied to the entire assembly from all directions or through the substrate, or locally to the solder paste or the component leads. It can be applied by radiation, conduction, convection or induction. Some of the more popular methods are considered in the following sections.

7.3.4.1 CONDUCTION METHODS FROM BELOW

Heat can be transferred to an assembly by conduction from a source in the form of a hot solid or liquid. The former technique is used with thermally conductive substrates such as alumina or beryllia and relies on a suitably flat base to enable heat to be transferred from the hot plate through the substrate to the solder paste on the top surface (see Figure 7.3). The main problem is to ensure uniformity of heat transfer over the substrate area. As the component leads are at a lower temperature than that of the substrate, it is difficult to time the exposure of the assembly to the hotplate, such that those components with a high thermal demand are adequately soldered, while smaller components are not over 'cooked'. One method used to solve these heat transfer differences is to place the substrates on an intermediate PTFE belt which moves over a series of hotplates at

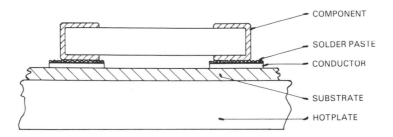

Fig. 7.3 Reflow soldering by the conduction of heat through the bottom surface of the substrate from a hotplate.

different temperatures, thus creating a heating-cooling cycle, e.g., preheat, reflow, cooling.

Heating from below can be used for epoxy laminates and one method employs a wave of hot oil in a wave soldering machine. A long heating time is required (20-30s), so a low conveyor belt speed is necessary. Selection of the most suitable oil is dependent on the substrate material used as the method demands a close molecular contact to ensure good heat transfer without permitting the flooding of oil onto the top surface of the substrate.

7.3.4.2 HEATING FROM ABOVE

Three methods of heating from above, used in conjunction with solder pastes, employ infra-red radiation, hot gas, and laser beams. The latter two are examples of localised heating while infra-red can be focused local heating or diffused generalised heating. A fourth method not normally employed with solder paste is described in Section 7.3.4.4.

7.3.4.2.1 Infra-red Soldering

Infra-red sources can be tungsten filament lamps, nichrome alloy quartz tubes, or panels operating on a secondary emission principle. The resulting wavelengths will vary from 1 to 6 microns depending on the type of source[4]. Their advantages are low capital and running costs, and both heat and time are fully adjustable. To be able to achieve a controlled pre-heat and cooling down situation, and so prevent over-heating, the substrates are introduced to the heat sources on a conveyor belt. The main problem is that the ultimate temperature achieved by the various components and the substrate is dependent on the thermal mass, conductivity, absorption and reflection properties of the materials involved and on the wavelength of the radiation. To some extent this problem can be reduced by using pre-heating plates underneath the conveyor. These raise the substrate temperature to below the solder melting point leaving the infra-red radiation only to raise the temperature the remaining amount required for the solder to flow. The solder paste is a good absorber of radiation (40% absorbed with radiation wavelengths of 1 micron)[2], while the molten solder strongly reflects I-R radiation. Because black is a good absorption colour, it may be necessary to shade black plastic encapsulated devices. Alternatively, some irregular or large shaped components may cause shadowing of an area that needs to be exposed to the radiation. These problems can be reduced by using non colour-selective diffused

I-R energy sources. Area source emitters operating at middle to far infra-red frequencies also heat the air[5] and this is claimed to greatly reduce shadowing effects due to the moving hot air. Because infra-red may need careful adjustment for a particular assembly, it is perhaps best used where there is a limited variety of production throughput.

7.3.4.2.2 Laser Soldering

Laser beam soldering of individual joints sequentially is quite feasible, a total energy of about 1 Joule being required. A CO_2 laser (10·6 μm wavelength) can complete a joint in 200-400 ms. Alternatively, a pulsed Nd-YAG laser (1·06 μm) will accomplish the heating[2] in about 4 ms. The main problems are to ensure that all points are in the line of sight of the laser, and, for production qualities, the laser beam must be capable of being rapidly and accurately positioned, for example, by microprocessor control. Because the heating up time is very rapid, it is necessary to choose the solder paste wisely, otherwise the boiling of the flux solvent may be so violent that solder or components are blown away.

The main advantages stem from the very localised area of the laser beam spot (0·3 to 1·5 mm diameter), concentration of heat onto the joint, and the short soldering time. Solder bridging is unlikely, board delamination is avoided and very small or thermally sensitive components are unlikely to be damaged. It is therefore one reflow method applicable to mix print applications.

7.3.4.2.3 Hot Gas Soldering

Gas can be passed over an electrical heating element and subsequently blown onto the component joints to be heated. The gas is usually air but nitrogen or a nitrogen hydrogen mixture can also be used. In order to prevent the gas disturbing the positioning of the components or overheating them by being too hot, the velocity and temperature are limited and the heating process can be slow. The method has advantages in that the components cannot be mechanically damaged and heat can easily reach the parts requiring to be heated. It is often used for the soldering of multipin components such as chip carriers, or for rework.

7.3.4.3 VAPOUR PHASE SOLDERING

Vapour Phase Soldering is an example of all-round heating and is also called condensation soldering. It utilises the latent heat of vaporisation of a saturated vapour which is given up when condensing on the relatively cool electronic assembly introduced into the vapour. Because the temperature of the vapour is closely defined, and because once the products reach this temperature, condensation ceases and no further heating occurs, it is fundamentally a very controllable process and needs no separate temperature controllers. It is, of course, necessary to limit the time duration of immersion in the vapour to prevent damage to components.

Typical boiling points include:[6,7]

ISC Ltd—	PP11 at 215°C
3M Co.—	Fluorinert FC43 at 174°C
	FC70 at 215°C
	FC71 at 253°C

Montedison Co.—   LS range 220-235°C
                                   HS range 250-265°C

Figure 7.4 illustrates the simplest form that a vapour phase solder system can take. In practice the primary liquid used to generate the vapour is extremely expensive. To minimise vapour losses, a secondary vapour blanket of a less expensive and lower boiling point substance (such as trichlorotrifluoroethane, boiling point 47°C) is floated on top of the primary vapour. This is not only an economic necessity, but also ensures that possible harmful decomposition products, which can be generated, are not released to the atmosphere. Products such as hydrofluoric acid (HF) and perfluoroiso-butylene (PFIB) can be kept well below maximum acceptable levels by avoiding overheating the vapour and employing adequate ventilation.

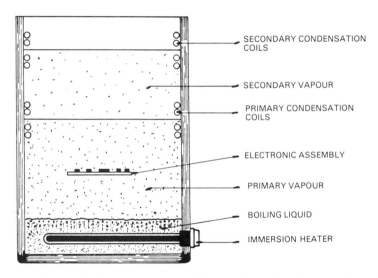

SECONDARY CONDENSATION COILS

SECONDARY VAPOUR

PRIMARY CONDENSATION COILS

ELECTRONIC ASSEMBLY

PRIMARY VAPOUR

BOILING LIQUID

IMMERSION HEATER

Fig. 7.4 A typical batch dual vapour system used for the all-round application of heat for reflow soldering.

It is good practice to pre-heat the assembly and its jig to a minimum of 55°C prior to insertion in order to prevent the blanket vapour condensing. Condensation can cause the blanket to collapse and thus allow the expensive vapour below to escape and may also cause dissolution of the flux.[8]

An alternative to the vertical batch machines is the in-line system illustrated in Figure 7.5, in which additional cooling zones are used to limit the vapour losses. The advantage is that products can be continuously processed on a conveyor belt. One possible drawback is that the resulting equipment becomes quite large, i.e., 15 or 16 feet long.[9,10]

The vapour phase soldering method offers very uniform heating irrespective of the thermal conductivity, size and shape of the components. It is therefore particularly attractive for soldering complicated products and where the production throughput is constantly changing from one design to another. A further advantage is that soldering is carried out in an inert atmosphere without fear of oxidation and at a lower temperature than many other methods. The process is a particularly clean method, but if flux residues are still considered

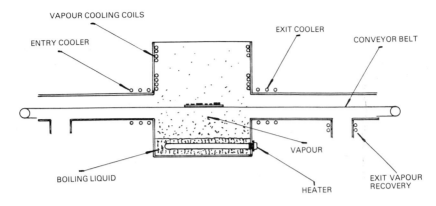

Fig. 7.5 The main elements of an in-line vapour phase reflow soldering machine.

objectionable, and must be washed away, it is easier because of the lower temperatures and lack of air. Because of its all-round heat application, vapour phase soldering is the only one of the four soldering processes which, in principle, enables both sides of a substrate to be soldered simultaneously. In practice it may be necessary to glue the components on the underside! It cannot be used as the only soldering process in mix print applications because of the damage caused to temperature sensitive bodies of some leaded components.

7.3.4.4 RESISTANCE HEATING

The previously described reflow soldering methods have used a solder paste. It is possible to reflow using solder in the form of a hard metallic coating which has been previously electro-deposited or hot dipped onto the substrate conductor surfaces. A method particularly suitable for pre-coated conductor surfaces is resistance heating which can be considered as similar to the application of an electric soldering iron, but using a relatively simple machine to offer greater control of the soldering operation. With resistance heating an electric current is passed through a specially shaped electrode or, alternatively, through two closely adjacent electrodes and also through the part to be soldered (see Figure 7.6). In

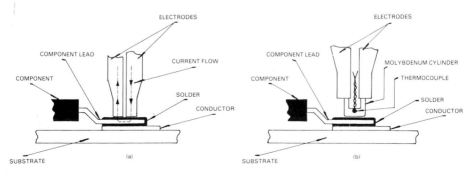

Fig. 7.6 Resistance heating for the local application of heat to a component's lead. In (a) the current passes through two adjacent electrodes, and also the lead to be soldered. In (b) the molybdenum cylindrical 'bit' is heated.

the first case the electrode is heated by the current passing through it, creating a so-called 'Hot Tip', and the heat transferred by thermal conduction to the joint to be soldered. In the second case the temperature of the part is directly raised due to its own resistance and its contact resistance with the two electrodes. By suitably adjusting the pressure of the electrodes on the joint and the current amplitude profile, the amount of heat transferred can be accurately controlled. Heating is very local and the application time can be short[3], typically 4s. It is, however, essentially a method for relatively low production quantities as the number of joints that can be soldered at a time is limited, and the total time taken, including positioning the component, may be 30s or more; also the thickness of solder on the substrate needs to be accurately controlled and to be quite thick (typically 20 μm). In practice, this may be best achieved by hot-tinning by passing the boards over a wave-soldering machine. Alternatively, they can undergo a hot air levelling process, in which the fluxed boards are dipped vertically into a solder bath and, on withdrawal, are exposed to strong blasts of hot air to remove excess solder. Roller tinned boards will generally have too uneven a surface and too little solder.[11] Although it is claimed that solder paste can be used rather than a pre solder-coated substrate, problems of spattering and solder balling make this less attractive owing to the higher temperatures reached by the electrodes.

Multiple soldering of joints is possible with resistance soldering using a 'heated collet'. Although more often used for rework (removing a previously soldered component and resoldering a new component), it can also be used to attach chip carriers and similar multiple way components onto a board which has been wave soldered. The collet applies heat simultaneously to all four sides of a chip carrier while applying downward pressure onto the component to hold it in place on the substrate. The heating element inside the collet is then switched off but the pressure is maintained until the solder solidifies.

7.3.4.5   SUMMARY OF REFLOW SOLDERING

The amount of solder used with reflow soldering methods can be accurately controlled and can, with some accuracy—depending on the method employed—be deposited where it is wanted, and equally, not where it could be an embarrassment in causing short circuits. It is therefore excellent for use with SMDs and fine line conductors and spacings. An added advantage with solder cream reflow methods is the natural phenomenon due to surface tension: as the solder melts, small components float on the molten solder and minor positional errors can be removed by the surface tension forces seeking equilibrium. This effect should, however, be treated with some caution because, if heating of the solder is uneven and one end melts before the other, the surface tension can cause displacement, even to the extent of forcing some small components, such as chip resistors, to stand on one end. This is sometimes referred to as the Tombstone effect, for obvious visual reasons, and can be reduced by ensuring that the solder cream is free of solvents before reflow soldering, that the amount of metallisation under the component contact is at least 0·3 mm and the device covers the solder land by greater than 0·3 mm. There is some evidence that this effect is most prevalent with vapour phase reflow soldering. Extending the pad length under the component body provides a torque counteracting the one which can cause the tombstoning effect.

The amount of solder available with solder paste is insufficient to solder leaded components satisfactorily. The only satisfactory method to ensure sufficient solder with reflow methods is to place a solder preformed ring around each lead. This can be tedious, particularly with large scale production.

Reflow soldering is ideal for substrates which are 100% SMDs, or where the finest line spacings are required on one side of a two-sided assembly. With a mixture of SMD and wire and tagged components inserted through holes in the substrate, alternative or additional methods are generally necessary. These are described in the next sections.

7.4 GLUE AND SOLDER ATTACHMENT

There are two fundamental differences between these techniques compared with reflow soldering. The SMD is now fixed in position prior to soldering and the solder is applied in the molten state to the underside of the substrate. These two changes are of course complementary—it would be difficult to turn the substrate upside down in order to solder it without first fixing the components in position. The usual method of fixing is by applying an adhesive between the body of the component and the substrate. Criteria for selecting, dispensing and curing the adhesive are considered in the following sections. There is also some discussion in the preceding chapter on the machine dispensing of the glue. The following sections will describe some of the possible soldering methods that can be used once components have been successfully attached to the substrate.

7.4.1 Selection of the Adhesive

The list of requirements to be met by the adhesive is surprisingly long. They include:

(i) The adhesive must be non-conducting electrically.
(ii) The dispensing of the correct amount to bridge the gap between substrate and components must be controllable and easily achieved.
(iii) The viscosity must be such that a blob of adhesive does not slump and flow near to the conductor pads and hence inhibit the following soldering process.
(iv) The glue must be sufficiently 'tacky' before being cured to retain the components in position.
(v) Preferably, it must have a short cure time.
(vi) Pot life must be sufficiently long so that machine-down time for cleaning out and refilling of the reservoir is low.
(vii) The cured adhesive must have adequate strength to retain the components during subsequent handling prior to soldering, and be able to withstand direct immersion in molten solder without outgassing or losing its holding properties.
(viii) These parameters must be met over the ambient temperature range likely to be experienced under normal factory or laboratory conditions.

In practice the above can be met by two-part or one-part heat cured epoxies, or U-V curable acrylates or epoxies. A standard industrial oven can be used to cure the epoxy adhesive, although a belt fed tunnel oven is preferable for higher production volume. The U-V cured adhesives are attractive as curing times can be short and pot life is long, but it is difficult to ensure that the U-V light can reach the adhesive under all the opaque components. This problem is best overcome by using two dots of adhesive on either side of the component, each dot only partly

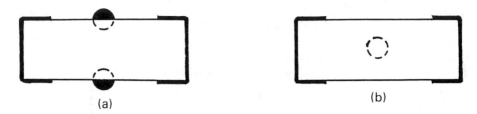

Fig. 7.7   Adhesive dot patterns (a) U-V cured, (b) Heat cured.

obscured by the body of the component (Figure 7.7). In general, the holding strength of U-V cured adhesives is less than that of heat cured ones, necessitating larger or an increased number of dots. They may also require a heat cure in addition to exposure to U-V.

7.4.2   Glue Dispensing

Glue can be placed either on the underside of the component or on the surface of the substrate. It can be dispensed by hand from a syringe or screen printed directly on the substrate. Neither of these can be recommended for production, the former being slow and not easily controllable and the latter, although quick and simple, cannot easily achieve the correct height of adhesive. The more usual techniques are either to pressure dispense a measured blob by machine onto the substrate, or to use a pin transfer system.[12] The latter method is illustrated in Figure 7.8 and shows glue being deposited onto the substrate. The Philips range

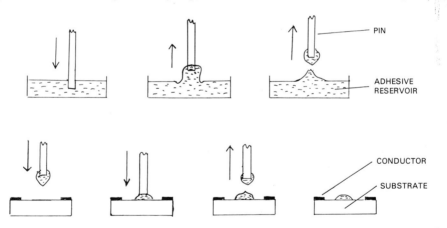

Fig. 7.8   Pin-transfer. Adhesive is removed from its reservoir, carried as a droplet on the tip of the pin and finally transferred to the substrate between the contact pads.

of MCM I and II chip placement machines use a similar approach, but place the glue onto the underside of the component rather than the substrate. The amount of glue required will depend on the gap between the underside of the body of the component and the substrate surface. In the case of resistors and capacitors, the adhesive must bridge the gap consisting of the sum of the thicknesses of the substrate conductor track and the metallisation on the underside of the component. A more difficult task may present itself with legged components such as SOT23 transistors. The adhesive must now bridge the substrate conductor thickness, plus the leg height of the component. Low-profile versions of such devices can now be purchased with reduced leg height to ease this problem. Where the distance is too large, or variation in the metallisation thickness of the substrate conductors will cause problems, a solution is to run dummy conductor tracks under the component (Figure 7.9). Small devices can be glued with one

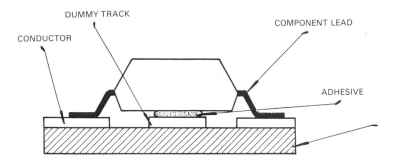

Fig. 7.9 A dummy track placed under the body of a component reduces the distance to be bridged by the adhesive.

drop of adhesive while larger components will require two or more. As an example—in the SO range of encapsulations for integrated circuits, an SO-16 can be bonded with one droplet while SO-28 requires two.[3] The size of droplet used will clearly depend on the glue chosen, the method of application and the room available on the substrate. It typically could be 1-2 cubic mm and have a diameter of 2-4 mm. Its height will be about 20% of its diameter.

7.4.3 Handling after Adhesive Dispensing

Once the adhesive has been dispensed and the components have been loaded into their correct positions on the substrate, it is necessary to cure the adhesive. To achieve this successfully will entail removing the substrate loaded with components from the assembly machine or work position and transporting it to an ultra-violet source in order to polymerise the adhesive, and/or to a heat source to cure it. This transportation must be achieved without any of the components moving. Experiments carried out deliberately applying shocks to assemblies with uncured epoxy adhesives proved that most components can withstand considerable mechanical shock, and show that no component movement should occur during normal handling in a production area. The most likely components to be affected are the larger ones such as VSO-40 ICs and chip carriers, and these are the ones to be checked when introducing a new assembly into production. Similar experiments to measure the shear strengths after curing on glass epoxy printed circuit boards showed that the bond between component body and substrate or dummy track was so strong that it is the bond between the solder resist and the substrate that will first collapse. No discernible differences in adhesive bond strength are detected before and after flow soldering although, once soldered, this is of little importance.

It was mentioned in Section 7.4.1 that suitable adhesives available are either U-V cured or thermoset. Probably the easiest method is to use a one-part epoxy and cure in an oven. The curing time will depend on the epoxy chosen and the

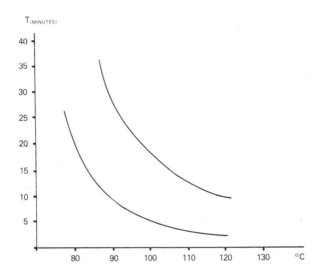

Fig. 7.10   Curing characteristics of two epoxy adhesives. Time to cure versus oven temperature.

temperature. Figure 7.10 shows the curing characteristic of two typical epoxies (the time axis excludes warm-up).

7.4.4   Machine Flow Solder Methods

The majority of printed circuit board assemblies using wire or tag terminated components are soldered on a conveyor belt system of machine soldering. In all cases there will be a loading position, followed by fluxing, pre-heating, soldering, cooling and unloading. A cleaning process may also be included. Preheating is a critical factor with SMD soldering due to the limited dwell time in the solder particularly with jet wave soldering. Ideally the substrate top side temperature should be about 100°C just prior to soldering. Soldering involves a meeting between molten solder and the surface of the board to be soldered.

This meeting together can be achieved by:

1   Dip Soldering—where the board is lowered onto a static bath of solder and then removed.
2   Drag Soldering—where the board is dragged horizontally across the surface of a static solder reservoir.
3   Wave Soldering—in which the board is dragged almost horizontally across the crest of a wave of flowing solder.

2 and 3 are the most popular modern methods. The one almost universally used to solder substrates using SMD or a mixture of SMD and wire terminated components is Wave Soldering, and this method will be discussed in greater detail. Before doing so, a brief résumé of dip and drag soldering follows.

7.4.4.1   DIP SOLDERING

The assembled board is lowered onto a solder bath at an angle between the horizontal, and about 30° to it, until breaking the solder surface when the angle is reduced to zero. Gas bubbles can better escape if the assembly is so angled on entry. A reversal of this procedure when withdrawing the substrate allows excess solder to escape and reduces the chance of bridging. The main advantage of dip soldering is that the investment is low, and maintenance is simple. Without some automation the results are variable depending on the fallibility of the operator. Because solder bridging is difficult to prevent when spacings are close, and solder cannot reliably be introduced into plated-through holes on double-sided substrates, this method has limited use with SMAs.

7.4.4.2 DRAG SOLDERING

The substrate is lowered onto a solder bath at an angle of about 15° to the horizontal and, as it contacts the solder, this angle is reduced to zero. This again allows gas bubbles to escape. The assembly is then dragged a fixed horizontal distance along the surface of the solder and then lifted away. The surface of the solder is previously skimmed to clean away any dross immediately before soldering each substrate. The process is simple, easy to maintain and free of human interference and variability. The level of solder is critical. The heat transfer, although better than dip soldering, is not regarded by some authorities as sufficient to guarantee good soldering of plated-through holes used in double-sided or multilayer conductor printed circuit boards. Furthermore, because drag soldering is essentially a horizontal flow process, it has significant limitations for SMDs. This will be emphasised in the next section.

7.4.4.3 WAVE SOLDERING

To achieve a correctly soldered SMD, the solder must reach the conductor pad of the substrate and the metallisation of the component[13] and the flux gases generated must be able to escape. With leaded components the solder can flow easily around the lead and gases can escape up the insertion holes. The majority of SMDs are placed directly in contact with the substrate and solder must now flow around the body of the device. One can have situations where the body of a component causes a shadow effect when using a conventional solder wave, so that if the direction of solder flow is as shown in Figure 7.11(a), solder, due to its surface tension, will not reach the trailing joint. Figure 7.11(b) illustrates a similar phenomenon where the leading component blocks the flow of solder to the next one. With the high component densities of SMA, this is a significant problem. The solution to this is to direct the flow of solder upwards against the underside of the substrate. To achieve this requires different waves from those previously used for leaded components. Two basic types are available—Single Jet or Double Wave.

7.4.4.4 SINGLE JET WAVE

In the jet wave machine, solder is impelled upwards through a horizontal orifice producing a thin stream of solder with a parabolic shape. The solder in the waves moves very fast (2m/s) and is made to circulate completely around each component so ensuring that all contacts are wetted. The high velocity wave also tends to scrub the substrate surface displacing any excess flux and removing the gases produced[14]. An interesting feature of the Kirsten Jet Soldering Machine (Kirsten Kabeltechnik AG, Switzerland) is that the wave is electrodynamically

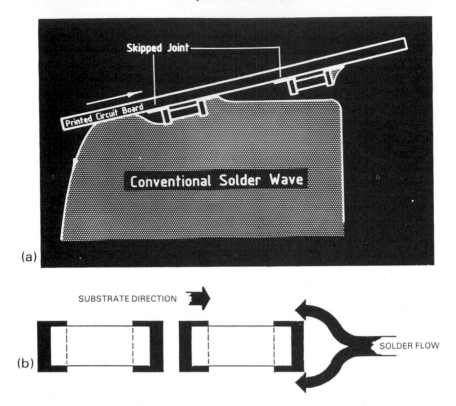

Fig. 7.11 (a) The surface tension of the wave prevents solder reaching the trailing joint of the component. (b) The leading component blocks the flow of solder to the following one.

generated—a current is passed through the solder at right angles to a surrounding magnetic field.

7.4.4.5 DOUBLE WAVE SYSTEMS

One problem of directing solder vertically upwards to ensure good wetting is that a large amount of solder is spread in all directions in a turbulent manner and this can introduce solder bridging. With a double-wave system the second wave provides the solder drainage and thus helps to prevent the incidence of short circuits. This may only partly succeed as the first wave may leave insufficient flux for a successful soldering operation by the second. A soldering oil or colophony can however be added as a thin film on the surface of the solder to reduce oxidation and decrease the surface tension. The second wave follows the more conventional form used for leaded components. Since the early fifties, various types of wave have been developed[15], one of the most popular now being the Lambda Wave (Electrovert patent) (Figure 7.12). The nozzle design enables a 4 in. contact length between the substrate and the solder to be achieved with a controlled exit zone where there is a low angle between substrate and solder, and zero relative velocity between substrate and solder flow, together with fast conveyor throughput speeds. A number of companies now supply dual wave machines for use with SMA or mixed technology assemblies. In all cases the

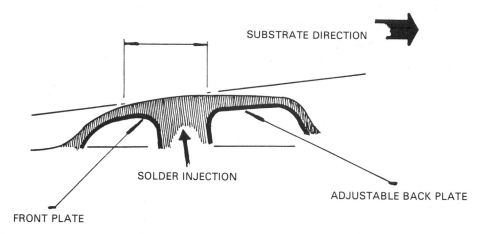

SUBSTRATE DIRECTION

SOLDER INJECTION

ADJUSTABLE BACK PLATE

FRONT PLATE

Fig. 7.12 Standard Lambda Wave.

primary wave is narrow and turbulent, while the second one is much wider and generally flows counter to the direction of the conveyor. Where the two waves are generated by separate pumps, the primary wave need only be operated when SMDs are being used on the substrate. Figure 7.13 illustrates a typical dual wave

Fig. 7.13 A typical Dual Wave System.

system. The optimum distance between the two waves should be between 2 and 3 inches[16]. The substrate is deeper in the primary wave than the secondary wave. The smooth second wave, as well as removing excess solder, allows sufficient time for capillary action to take solder into the plated-through holes and completes the wetting action. A typical conveyor speed is 6 ft/min giving soldering times of 0·5 to 0·8s in the turbulent wave and 2 or 3 in the smooth one.

7.4.4.6 HOT AIR KNIVES

The ever-increasing demand for higher component packing densities enhances the likelihood of solder bridging. Indeed, with leaded components bridging is necessary between the load and the annular ring of conductor surrounding its insertion hole. The requirement is to eliminate unwanted solder bridging. Molten solder exhibits a much greater tenacity in adhering to a properly wetted interface than to non-wetted surfaces, i.e., to the conductor rather than the substrate or solder resist surfaces in between. This property can be exploited by using an 'Airknife', which is a fine jet stream of high velocity hot air which is blown onto the still molten soldered surface immediately after wave soldering. It is adjusted in distance, air temperature (approximately 390°C), pressure and angle relative to the substrate (typically 45°) until it removes any bridges without affecting the solder joints. This method illustrated in Figure 7.14 was first developed in 1977 by Hollis Engineering[17] and Digital[18] for printed circuit boards using wire terminated components. One problem that can be encountered is that solder can be blown onto the top surface of the substrate through any holes in it unless the knife is suitably adjusted.

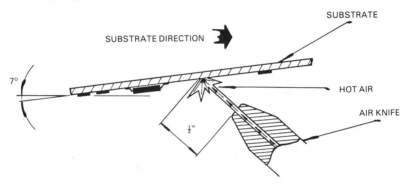

Fig. 7.14 Air Knife operating on the underside of the substrate after wave soldering.

7.4.4.7 ORIENTATION OF COMPONENTS

The substrate layout design including the required footprints has been discussed in Chapter 5. It is necessary to emphasise that the direction of flow through the solder wave will affect the soldering of multi-terminal components such as SO and chip carrier ICs. It is easier to achieve good solder penetration if an SO pack is laid parallel to the direction of flow (Figure 7.15). Transverse

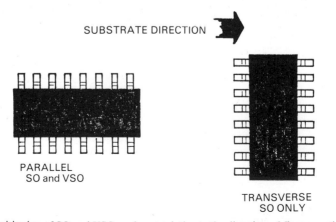

Fig. 7.15 Positioning of SO and VSO packages relative to the direction of flow over the solder wave.

positioning may not be possible with the closer spaced VSO-40 package, and, with the 0·05 in. pin spaced SO packs, is only possible if the conveyor speed is slow enough and the solder lands are sufficiently narrow. It may be found desirable to lay chip carriers diagonally and pass the substrate over the solder wave twice, the second time rotating the substrate through 90° relative to the first pass, although, with care, this second pass can be eliminated.

7.4.4.8 SUMMARY OF FLOW SOLDER METHODS

While reflow soldering is desirable for the closest conductor spacings, the majority of surface mounted assemblies can be satisfactorily soldered using wave flow solder methods, and is preferable if the designer demands a mixed print technology, i.e., surface mounted and conventional leaded components. A double wave machine is recommended in order to achieve good soldering of the chip components and leads inserted through substrate holes, without bridging. It is necessary with flow methods to orientate the surface mounted components correctly with respect to the direction of travel of the substrate over the solder wave. This is particularly so with the closer spacings and multiple terminations of IC encapsulations such as chip carriers and VSO packages. One of the interesting aspects of flow soldering is the occasional creation of a convex solder profile of the joints on the trailing edge of a chip component, particularly when using a Jet Soldering Wave (Figure 7.16). Historically, Quality inspectors would have

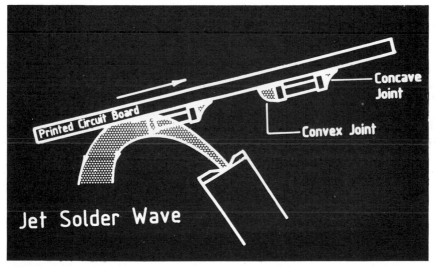

Fig. 7.16 Jet Solder Wave, while ensuring all solder joints are covered, can produce a convex profile on the trailing edge.

regarded any solder joint that was not concave as a sign of poor wetting. Examination of micrograph sections through such convex joints indicates perfect wetting at both the component and substrate surfaces (Figure 7.17).

7.5 HAND SOLDERING

It is possible to use manual methods of soldering SMDs using a light hand-held electric soldering iron, but the small size of the components makes it difficult to

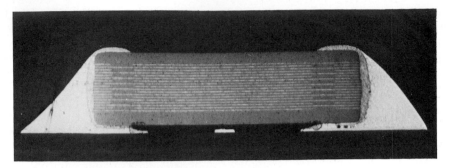

Fig. 7.17 Sections through convex solder profile joint showing good wetting at both component and substrate surfaces.

position the component accurately and to maintain the position until the solder has solidified. Care is also necessary to ensure the component's encapsulation is not damaged by inadvertently placing the hot iron surface onto the body. In practice this method is restricted to producing limited laboratory samples or for repair work on the production line. The normal rules of good hand soldering apply, including the use of an iron having a controlled temperature, and which is well earthed if electrostatically sensitive devices are being used. The application of sufficient heat for a short time is preferable to too low a temperature iron for a longer period.

7.6   CONDUCTIVE GLUES

Conductive epoxy based glues have been used for back bonding unencapsulated silicon chips to ceramic substrates for many years. The material is sufficiently compliant to prevent problems of cracking when subjected to thermal cycling. It is quite feasible to screen print such an adhesive onto a substrate to provide electrical connections to all of the components, thereby eliminating the need for soldering and all the associated processes involved, such as fluxing, heating and cleaning. Generally, two epoxies will be employed. A silver-loaded electrically conductive one is printed, using a 200 mesh screen with a 0·0018 inch emulsion.[18] A non-conductive epoxy is deposited to provide added mechanical strength between the underside of the component body and the substrate. The epoxies require curing at a temperature[20] of about 150°C and such temperatures can be damaging to some types of component when applied for the period of time required by the epoxy—typically 5 to 30 minutes. The epoxy can be cured in an oven, infra-red tunnel or with vapour phase equipment (1 minute at 210°C for certain epoxies). The electrical conductivity of the adhesive is poor compared with that of solder and thus will be inappropriate for high frequency and high speed applications. Advantages claimed, however, include the elimination of the problems of leaching of component metallisation and easier rework after fault finding as the faulty component can be readily removed once its terminations have been heated sufficiently to break the glue's bond strength.[21]

7.7   POST CLEANING

The importance of post-soldering cleaning will depend on the intended application of the product. Military requirements will be more demanding than consumer applications. The main arguments for cleaning are either based on the potential long-term degradation, due to chemical attack from materials used in the flux or generated during the soldering process, or shorter-term physical problems due to the presence of flux residues preventing good electrical contact of testing pins or edge connectors, and the difficulty of inserting boards into racking. There can also be the purely visual argument that washed boards appear more attractive.

The need for washing will depend upon the choice of flux used in the soldering process. A successful flux removes surface oxides and prevents re-oxidisation, encourages the formation of an alloy between the solder and the metal surfaces of the component and substrate and helps in the transfer of heat to the joint area. Fluxes are based on either rosin from natural pine sap or so-called water soluble compounds. Either can be activated by the addition of chemicals aimed at cleaning the metal surfaces. Water soluble fluxes are generally used to provide high fluxing activity and consequently their residues are more corrosive. The same fluxes can be used for wave soldering or in the solder paste for reflow. Successful fluxing using a paste generally requires a more active flux and may increase the need for intensive cleaning afterwards. Combatting this, however, can be some self cleaning effect of using a reflow solder process such as vapour phase. Highly activated fluxes demand thorough washing, while many rosin-based fluxes can be left on the substrate. If a water-soluble flux is used, the solvent employed in the cleaning process can be water, which has the advantage of being cheap, but then introduces the problem of drying the assembly afterwards. The alternative is to use one of the many available proprietary fluorocarbon solvents.

The special problems of cleaning assemblies using SMDs is that, by their very nature, the devices offer a very low clearance between the component body and the substrate surface. It is therefore difficult for the cleaning fluid to access all the areas that need to be cleaned. Pressure washing or ultrasonic cleaning may become more necessary to overcome the problem. In reported experiments[22] using chip resistors and capacitors, SOT23s, SO ICs and 44-pin plastic leaded chip carriers, cleaning under SO ICs was markedly more difficult than equivalent control board areas without SO ICs, while little difference was observed with the chip carriers. This could well be due to the very small clearance (0·008-0·009 in.) between the underside of the SO devices and the FR4 laminate.

The cleanliness of an assembly is generally checked by visual inspection or ionic contamination measurement equipment. Unfortunately neither can guarantee that all contaminants have been removed from between the surface mounted components and the substrate.

REFERENCES

1 Roos-Kozel, B., 'Surface Mount Technology,' ISHM Technical Monograph, Series 6984-002, Chapter 5, 'Solder Pastes'.
2 Klein-Wassink, R. J., 'Soldering in Electronics', Chapter 10, 'Reflow Soldering', Electrochemical Publications, Ayr, Scotland.
3 Mullard Technical Publication 'Guidelines for Surface Mounting of ICs in SO and VSO Packages', Mullard Ltd, January (1985).
4 Dow, S. J., 'The Use of Radiant Infra-red in Soldering Surface Mounted Devices to Printed Circuit Boards', *Brazing & Soldering*, No. 8, p. 16, Spring (1985).
5 Dow, S. J., 'Using Radiant Infra-red to Solder Surface Mounted Devices', *Electronic Production*, p. 19, September (1985).
6 McMillan, I. D., 'Vapour Phase vs Infra-red Heating for In-line Solder Reflow,' *Electronic Production*, April (1984).

7   3M 'Fluorinert' Fluids for Vapour Phase Heating, ICI Solvents Marketing Dept.
8   Hey, Dr D., 'Entering the Vapour Phase', *Electronics Manufacture & Test*, January (1984).
9   In-line Vapour Phase System. Leaflet C48, Hedinair Ltd.
10 Vapour Phase Process. Bath and In-line Production Systems Pamphlet, ICI and HTC.
11 Roberts, L., 'Hot Tips Offer Soldering Success,' *Electronics Manufacture & Test*, December (1984).
12 Philips Export BV, *Electronic Components & Applications*, **Vol. 6**, No. 2, p. 82 (1984).
13 Guidelines Surface Mounted Assembly, Publication No. 12845401, Mullard Ltd, August (1984).
14 Walton, A., 'A Soldering Machine for Surface-mounted Components,' *Electronic Production*, February (1983).
15 Carr, R., 'Solder Waves—A Review,' *Electronic Production*, October (1984).
16 Owen, W. and Rahn, A., 'Waving Goodbye to Solder Skips,' *Electronic Manufacture & Test*, June (1984).
17 'Air Knife Cuts Through Bridge Formations,' Electrautom and Hollis, *Electronic Manufacture & Test*, May (1984).
18 Lambert, L., 'Airknives Aid Solder Defect Control,' *Electronic Production*, May (1984).
19 Kulesza, F. and Estes, R., 'Conductive Epoxy Solves Surface Mount Problems', *Electronic Products*, March 5 (1984).
20 Epo-tek 'H' series epoxies, Data sheet, Epoxy Technology Inc.
21 Freakes, A., *Electronic Engineering*, p. 156, September (1984).
22 Comerford, F., 'Putting Cleaners to the SMC Test,' *Electronic Manufacture & Test*, March (1985).

Chapter 8

POST ATTACHMENT PROCESSES

J. F. PAWLING and S. J. MULLETT

Mullard Mitcham, Mitcham, Surrey, England

8.1 INTRODUCTION

Once the construction of an electronic assembly is completed, including the removal of any residues such as solder flux used in preceding processes, it is necessary to confirm that a good one has been manufactured. The product must eventually be accepted visually, mechanically and electrically in its intended application. There are few markets where this acceptance testing of a product is left to the final user, and typically visual inspection and electrical test will be performed several times during the manufacturing process. The following sections of this chapter will separately consider Inspection, Electrical Testing, Fault Finding and Repair. The chapter will conclude by discussing the possible final post-attachment processes of conformal lacquering and encapsulation of the electronic assembly.

8.2 INSPECTION

Inspection of a component or an assembly is most often carried out visually by a human operator, although increasingly attempts are being made to machine inspect in order to eliminate many of the variables of human inspection brought

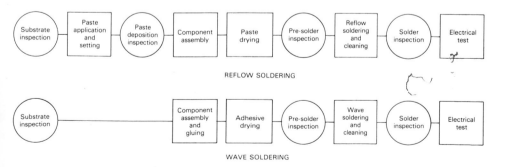

Fig. 8.1 Suggested inspection positions for typical manufacturing lines using reflow and wave soldering processes.

about by boredom, fatigue, and possibly ignorance. Inspection can be carried out at a number of points in manufacture. Figure 8.1 suggests positions for both reflow and wave soldering manufacturing methods. The main objective must be to alert the production supervision of any deleterious change in the manufacturing processes so that quick remedial action can be taken and not be in order to select the good from the bad. Where production rates are high, this is most important as the consequences of not doing so sufficiently quickly will show in large stock piling of rejects. On the other hand, large production through-puts coincide with 100% automatic component placement and insertion where faults are less likely to be the random ones associated with manual placement by human operators.

Each inspection area will be considered separately, with its objectives, the faults likely to be encountered, the standards to be aimed for, and possible methods that can be employed.

8.2.1 Substrate Inspection

Poor substrate quality can cause significant financial loss once the assembly has been completed, with the additional cost of all the expensive components, so, to be sure, 100% inspection is necessary. Where the substrate is a bought-in component, inspection can be carried out by the supplier, the user, or both. This is an item that can be inspected by automatic optical or electrical methods. The consistently repetitive nature of the substrate pattern enables an optical comparison to be made between the substrate being inspected and a known good one. This can be carried out, for example, by using a camera to send video information to a computer, for the comparison to be made automatically, and any differences indicated. This method will show up faults in the X-Y plane such as open tracks, short circuits, reduced pad areas and track widths, and large pin holes. It may not indicate hair-line fractures and Z axis faults such as delaminated tracks.

An alternative to the optical comparison method is an electrical one. This is achieved by applying contact probes to the circuit pattern 'lands' and testing electrically between them for disconnections or short-circuits[1]. The jig containing the test probes can be built uniquely for each substrate pattern or a general purpose matrix of probes can be used which is then uniquely programmed for the particular circuit layout. The former approach is used when quantities are large. While the general purpose approach is more attractive for medium and small numbers, it demands that the circuit pattern corresponds to a standard matrix pitch, typically 2·54 mm.

8.2.2 Solder Paste Deposition

Inspection is aimed at checking the screen printing alignment and the paste thickness. In practice, a sampling inspection scheme is generally satisfactory. Poor alignment and excess solder paste may cause bridging while poor alignment and less than optimum amounts of paste may reduce the amount of solder below that necessary to achieve a good solder joint (Figure 8.2). A typically accepted criterion is that at least 75% of the solder pad should be covered by solder paste. It is possible to use the automatic optical comparison methods described above to carry out an initial sorting. Rejected substrates can then be inspected by an operator when some degree of evaluation opinion is required.

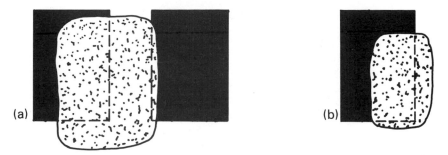

Fig. 8.2 (a) Poor alignment of solder paste and pad with excess solder causing bridging and (b) poor alignment and insufficient paste causing a suspect solder joint.

8.2.3 Pre-solder Inspection

This stage is aimed at checking the accuracy of component placement and even whether any components have been omitted entirely, and is usually a visual inspection by an operator. The degree of misplacement of a component that can be accepted will depend, for example, on the amount of overlap between the solder land on the substrate and the width of metallisation of the component. An acceptable misplacement for two adjacent chip capacitors is illustrated in Figure

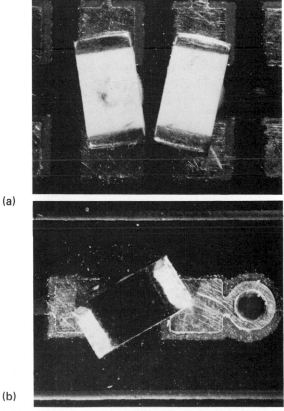

Fig. 8.3 Acceptable misplacement of two adjacent capacitors is shown in (a), while unacceptable misplacements for a chip resistor and SOT23 transistor are shown in (b) and (c).

(c)

Fig. 8.3 Acceptable misplacement of two adjacent capacitors is shown in (a), while unacceptable misplacements for a chip resistor and SOT23 transistor are shown in (b) and (c).

8.3(a) (provided the voltage breakdown requirement is met), while unacceptable misplacements for a chip resistor and an SOT23 transistor are shown in Figures 8.3(b) and (c). In practice, a larger overlap in width will be necessary with reflow soldering than with wave soldering. Where components have been fixed by adhesive prior to wave soldering, inspection is necessary to ensure that no contamination of the solder terminations has occurred due to the adhesives. It must be remembered that, at this and following inspection stages where active components such as ICs have been added, anti-static precautions must be taken to prevent the handling necessary with inspection procedures causing electrical failures of voltage-sensitive semiconductors.

With Surface Mounted components it may be possible to use automatic optical comparison methods to check the correct positioning or omission of components[2,3]. This approach is not possible using conventionally terminated components as correctly inserted devices can lie at varying positions, particularly if manually assembled (Figure 8.4). The general omission of colour coding with SMDs and their inability to flop over (unlike long leaded devices inserted through

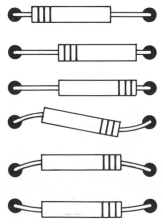

Fig. 8.4 Wire terminated components can take up a number of different positions—particularly when inserted by hand.

holes) mean that the visual differences between two loaded substrates will be small unless there is a real fault with one of them.

8.2.4 Solder Inspection

The high packing densities of SMA make the visual inspection of solder joints difficult. The partial obscuring of the joint by the component body, which occurs in many cases, exacerbates the problem (Figure 8.5). Some form of optical

Fig. 8.5 The component body makes visual inspection difficult by obscuring most of the solder joints.

magnification is essential, such as a microscope with zoom facilities, or preferably, to reduce operator fatigue, a video camera plus a high resolution screen. A requirement of good joints is that the part of the component metallisation not projecting past the edge of the solder land has a meniscus all along its length[4]. The shape of the solder joint will vary with the method of soldering. Wave soldering will deposit a large volume of solder and the joint of a surface mounted resistor or capacitor may be higher than the maximum thickness of the component and may therefore be convex rather than concave (illustrated in Figure 8.6 and referred to in the previous chapter, Section 7.4.4.8). With reflow soldering the amount of solder is generally less and the height of the joint will be below the height of the component, resulting in a concave meniscus.

Some components such as flat pack ICs have flat feet and only a relatively small amount of solder is needed to make good contact with the solder land. Others such as the SO IC are short, rigid and curved so that a wedge-shaped space is formed between the underside of the connectors and the solder land. A larger amount of solder is now desirable to fill this wedge.

The major solder faults to look for and eliminate include:

Insufficient solder and dewetting—no continuous meniscus between substrate pad and component (Figure 8.7).

Bridging—	solder connection between two separated conductors.
Cracking—	a crack in the solder which may become an open circuit with temperature cycling.
Flux residues—	may cause longer term corrosion or prevent later conformal coating.
Solder balls—	pieces of solder not firmly attached to solder lands and therefore likely to become detached and cause bridging.

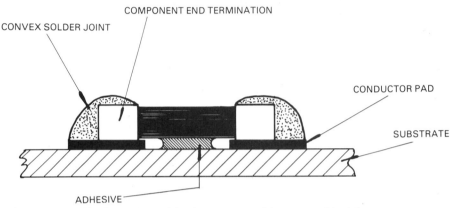

Fig. 8.6 Wave soldered component with convex solder joint.

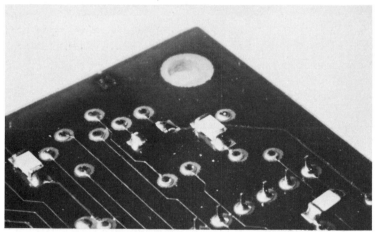

Fig. 8.7 Discontinuous solder line between component and solder pad.

Criteria for accepting or rejecting a solder joint, or component misplacements with respect to the substrate lands, will be set by a particular manufacturer and will depend upon the intended application, electrical requirements such as voltage breakdown, and the degree of control achieved over the manufacturing processes. Typical criteria set, for example, for wave soldered chip resistors and capacitors, are that the solder meniscus should be along its full length and have a minimum height of 0·4 mm, or one third of the height of the component's metallisation, if this is greater. For components with a few short legs such as SOT23 transistors, it is advisable to ensure that all of the feet are positioned within the boundary of the solder pads. On the edge is acceptable. Devices having large pin counts and capable of being visually inspected, such as SO and VSO packaged ICs, are usually required to have at least three quarters of the length of a foot secured to a pad. For encapsulations in which part of the lead is hidden, such as plastic leaded chip carriers, the quality of the solder joint must be judged by the visual quality and quantity of solder. A generally accepted minimum is a meniscus extending to a height equal to half the lead thickness, provided the entire solder land is wetted.

The extreme difficulty of achieving reliable 100% visual inspection of solder joints has led to much recent work on automatic methods of solder joint inspection. These include X-ray analysis, and evaluating the temperature rise behavioural differences between a good and bad joint, with laser scanning. X-ray inspection is reliable but can only operate slowly; the equipment is expensive and adequately trained operators are necessary for reasons of safety. With the latter system a laser beam is operated in a pulsed mode and heats up each individual joint. The resulting radiated heat is sensed by an infra-red detector and compared with the radiation pattern of a known good joint. The assumption is that a faulty joint, having cracks or pin holes, will be sufficiently different to be detected. The process can be automated and is quicker than X-ray techniques.

There is undoubtedly a need to avoid visual inspection of solder joints if at all possible. Where faults are common, then visual inspection is rewarding to both the inspector and the production management as many faults are being found relatively cheaply. However, with automatic assembly of surface mounted components and good process control, soldering defects should be well below 1000 ppm and, at these levels of incidence, visual inspection becomes tedious and difficult when coupled with the closer packing densities used with SMA. Figure 8.8 shows a not very complex assembly, yet it illustrates the difficulty of visual solder inspection. A board assembly containing a thousand or more joints will have a high statistical chance of failing electrical test, even if the solder joint failures are well below 1000 ppm. This gulf between the justifiable performance that can be expected from visual inspection, and the requirements of input quality

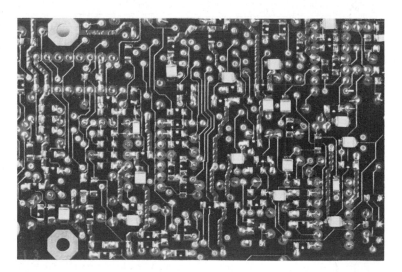

Fig. 8.8 Part of an assembly illustrating the difficulties of visual inspection of the soldering.

levels to electrical functional test, is generally bridged by in-circuit component testing.

8.3 ELECTRICAL TESTING

The completed electronic sub-assembly, consisting for example of a printed circuit board with its attendant components, must be functionally tested to confirm that it performs as the circuit designer intended. This can be accomplished by building the part into its final piece of equipment (a TV set, washing machine, car etc.), and checking that. This would be unusual, except for the cheapest consumer products and, generally, a purpose-built piece of test equipment is used to exercise the electronic sub-assembly in a manner as similar as possible to its final environment. This functional tester will be no different whether the assembly employs surface-mount techniques or not. However, because SMA offers opportunities for greater component packing densities and more complex systems in a given substrate area, visual inspection techniques are more difficult and, even with good process controls, significant numbers of failures can occur at functional test unless the board has been pre-screened electrically. This involves in-circuit component testing with its attendant increased capital investment, but with the bonus that faults found at this stage can be accurately pin-pointed and hence more easily rectified.

8.3.1 In-circuit Component Testing

This technique involves applying test stimuli such as a voltage between nodes of the electrical circuit, measuring the results and comparing against the expected values. With this method, every conductor can be checked for continuity, solder bridges can be detected, and every simple component such as resistors, capacitors, and semiconductor junctions can be measured and checked for correct polarity and value within the specified tolerance. To achieve this in a practical circuit, techniques have to be employed to prevent erroneous readings being obtained due, for example, to the effect of another lower valued resistor lying in parallel with the resistor under test, or to a nearby semiconductor junction being biassed forward by the applied stimulus. Methods such as 'Guarding' are used to overcome these problems.

To achieve the hundreds of tests involved within a sensibly economic time (i.e., a few seconds), the tests are sequentially carried out under the control of a computer previously loaded with the customised test programme for the assembly under test.

In practice, the test programme software can be generated in a number of different ways varying from a fully automatic process to employing a knowledgable electronics circuit engineer. The more sophisticated testers can offer the greatest assistance, and also the greater potential for being able to test the more complex components such as SSI and MSI. Where a fault is found, it can be flagged up by the tester, and indicated to the fault repairer, for him to read a printed hard copy report, or to interrogate a computer data base. To apply stimuli and measure the test parameters, access is required to all the internal circuit nodes using a so-called bed of nails, see Figure 8.9. In practice the 'nails' are spring-loaded probes, having a shaped tip able to cut through any remaining flux or tarnish on the conductor tracks and joints; slim to meet the demands of high packing densities, but strong enough not to buckle.

Fig. 8.9 An example of a 'bed of nails test' jig for a mixed print assembly.

One of the most prevalent problems with this form of testing is to ensure reliable electrical contact between the test probe and the substrate test point. This is achieved by bringing the two together pneumatically and/or mechanically, but additionally requires the correct shape of probe at the tip making contact with the substrate, sufficient spring contraction of the probe to take account of bowing of the board and varying heights of the test pad target and, above all, a clean substrate surface. Nothing is more frustrating for a fault repairer than to find that apparent high impedance faults signalled by the computer-aided in-circuit tester are faults in the test jig rather than a reject assembly.

With conventional wire terminated components inserted through holes in a printed circuit board, the conical solder joint which is formed provides a convenient landing place for such a test probe. With a mixed print technology, it may still be possible to use such a solder junction, but with surface mounted components this form of joint is not available. The sloping surface of a solder joint at the end of a surface mounted resistor or capacitor is not generally considered to be able to provide a reliable landing spot for a test probe. Chip carriers are especially hard to probe, since they present such a small target area because their leads hug closely to the base and then fold underneath. In practice, the solution may be to include specific test pads, but this is at the expense of surface area of substrate, thus limiting the achievable packing density (Figure 8.10). Such pads should be at least 0·15 in. away from the SMD[5]. An alternative is to access the device from the side[6]. A situation unique to SMA is that components can be assembled on both sides of the substrate. In order to check both, one must either pass the assembly through the tester twice, checking one side on each occasion, or achieve one pass by using a more sophisticated double-sided test jig[7]. A third possibility is to arrange that all the necessary nodal points are brought through to one side of the substrate using additional vias. This latter solution will again be at the expense of achieving the tightest packing density.

The application of surface mounted components to printed circuit boards has enabled much more complex circuits to be executed on one substrate. This

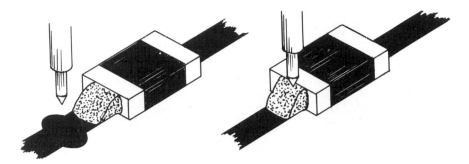

Fig. 8.10   Example of the layout for a test pad (a). This utilises more substrate area but avoids trying to land the test pin onto the solder joint (b).

consequently increases the required number of test nodes and has forced Automatic Test Equipment and Jig manufacturers to revise their thoughts on how many spring probes were required on a single test head. A maximum of one or two thousand was thought to be sufficient before SMA. The greater packing densities also demand that test pins can be placed much closer together, i.e., 0·05 in. or less. This results in more fragile probes, more expensive jigs and probes having a higher mean resistance[8].

A possible alternative strategy for the more complex substrates is to incorporate specifically within its design circuitry to exercise and test the remainder of the circuit. This is a technique often employed within a silicon integrated circuit.

8.4   FAULT FINDING AND REPAIR

The highest quality product results from a 'Right First Time' manufacturing approach. In practice, faulty products will occur and will require repairing. To ensure that repaired assemblies are also of a high quality, the minimum necessary repair processes should be carried out. Changing components unnecessarily is both expensive and likely to cause some deterioration in overall quality. Visual inspection, in-circuit component testing, and functional testing will all be stages in manufacture which will throw up possible rejects for repair. Visual inspection will indicate errors of component omission, poor soldering and unclean substrates, while in-circuit testers will provide a strong clue as to a particular component error, but may still demand some interpretation by the operator. For example, is a high impedance due to a wrong valued resistor being assembled, a poor solder joint or the tester probe making poor contact? Faults at functional test, if preceded by a successful pass-through on an in-circuit component tester, are likely to be IC faults such as marginal speed performance. If not preceded by an in-circuit tester, then the spectrum of possibilities is much greater.

The majority of faults are likely to be due to a component—either a faulty device, or a perfectly good device wrongly assembled. In both cases, the component has to be removed and replaced. The basic method is to melt the solder of all the component's joints at the same time and lift away from the substrate, clean the area and solder on a new device. With two or three terminal devices such as resistors, capacitors and discrete semiconductors, the removal can be accomplished using a small soldering iron and a pair of tweezers. Solder once melted can be removed by suction using a very simple manually operated pump or connecting to a more elaborately derived vacuum line, or by capillary action using a wick of copper braid[9]. If the component body has been glued to the substrate prior to wave soldering, the adhesive bond is usually easily broken by a combination of the heat of the desoldering process and a twisting action with the tweezers. This is dependent on previously selecting an adhesive that will fail during the removal operations[10]. The new device can be hand soldered into place. With multi-pinned integrated circuits the process used can be selected from a number of possibilities.

The connections to Dual-in-line devices such as SOICs can be heated simultaneously by a modified soldering iron head consisting of two parallel ridges of metal positioned to coincide with the two rows of solder joints. This is possible as the leads are more readily accessible than with chip carriers. With these latter devices, it is better to use either a stream of heated air (described in Chapter 7, Section 7.3.4.2.3), or a heated collet (Section 7.3.4.4).

The hot air approach can be combined with a secondary stream of cold air which prevents the heat spreading and possibly affecting other parts of the circuit. Before replacing the defective component with a new one, solder paste needs to be applied to the conductor footprint of the integrated circuit. This can be accomplished using a syringe, preferably combined with a dispensing machine. An ideal method with multi-terminated devices, such as chip carriers, is to use a mass dispensing tool which applies paste simultaneously to the whole footprint[11]. Because there is no physical contact, as there would be from an iron or collet, the self-alignment characteristic of the reflow process is maintained. On the other hand an advantage of using a heated collet is that the downward pressure ensures that all the component legs are in good contact with the pads. The heat source can be switched off and the solder allowed to cool and solidify before removing the applied pressure. With the SO IC packages, it is perfectly feasible to hand solder the replacement devices. The usual technique is to place the component in the correct position and solder tack two diagonally opposed leads to their conductor pads. The device's position is now fixed and the remaining terminations can be soldered and the original two retouched up if necessary. Other methods worth mentioning are to use a focused infra-red source of heat or a superheated inert gas such as nitrogen. The advantage of the latter method over hot air is the elimination of oxidation during soldering.

8.5 PROTECTIVE COATING AND ENCAPSULATIONS

Printed circuit assemblies often encounter hostile environments, such as moisture, salt spray, aggressive smokes and vapours. While certain applications restrict the environment to dry clean air and have little need for added protection and the consequential extra cost, there are many applications where some additional protection is necessary. Techniques well established for conventional electronic assemblies can equally well be applied to SMA. These can include solution-based conformal coatings (lacquers), solids-based conformal protection and complete encapsulation. With circuits based wholly on SMD there can be the advantage that higher temperature limits for the processing of the protection coating may be possible than with those assemblies using a mixture of SMD and conventional leaded components.

The following discussion will first consider various conformal coatings, and then encapsulation.

8.5.1   Conformal Coating

Conformal coating falls into two categories: solvent-based resin solutions, reactive and non-reactive, which give a thin $0 \cdot 1$ mm coating, and one hundred per cent solid systems which finally deposit a coat in the order of five to ten times as thick.

Conformal coatings provide protection from moisture and aggressive environments by total coverage of circuit board and components. This can generally extend the life of the assembly, provided the preceding cleaning techniques have been thorough and removed all forms of contaminants, fibres, rust particles, oils, greases, ionic and non-ionic residues, etc. To give circuit reliability, especially circuits of high impedance, experience has found that the coating performance is influenced both by the cleaning method and its control, and the degree of conformal coverage, i.e., freedom from bubbles, pin holes and poor drainage due to bulky components. Failure to obtain a clean surface will give rise to vesication, i.e., blister formation under the protective coating as a result of exposure to humidity cycling.

For these reasons, a simple coating will be effective as antisplash and general insulation, but for long-term service in extreme temperature and humidity environments, the coating has to be carefully selected and tested with circuits cleaned to very high standards. Comparative tests, to select a particular coating, should be carried out in controlled environments, with an applied voltage. Whichever method of application is chosen for the conformal coating, such as dipping or spraying techniques, the circuit will require terminals, interconnections and turning screws of potentiometers to be masked, which, unless carried out automatically, will add considerably to the cost. The use of more than one row of terminals per substrate can sometimes restrict the use of dipping techniques unless expensive masking techniques are employed.

8.5.1.1   DIPPING TECHNIQUES

Where dipping techniques are used[12] as a method of application for protective coatings to surface-mounted devices, the immersion rate is critical, particularly in the case of a close packed layout, in order to displace air surrounding, and under, the components. A soak, to enable air bubbles to be displaced, should be allowed before slow withdrawal. Any entrapped air can give rise to problems of blow-outs and bubbles if elevated drying or cure temperatures are used.

To eliminate pin holes, which can provide a point of entry for moisture, and control the reduction in thickness of the applied coating, when drying and stretching over sharp edges and protrusions, several coats are applied. The optimum number is determined by environmental testing. To ensure a consistent coverage, it is necessary to maintain the correct solution viscosity by the occasional addition of thinners. This is necessary to overcome solvent losses which will vary, both with the surface area of the bath and the throughput rate.

8.5.1.2   SPRAYING TECHNIQUES

Spray application provides the most expeditious method for coating circuits. To obtain a uniform coating over surface-mounted devices and to prevent bubbles and voids during the drying process it is necessary to control functions such as spray nozzle and pressure, the solid contents and the viscosity of the system. With conventional leaded component assemblies, spraying is required from various angles in order to ensure complete coverage. With SMA, which is closer to a two-dimensional assembly, this may be less critical, although several coats are required for the same reasons, as with dipping techniques. The spray technique will provide a bridging overcoat and is less likely to impregnate compared with dipping.

8.5.1.3 OTHER TECHNIQUES

Flow or curtain coating provides a gravity pulled wall of material through which the substrate is passed. This is another approach used for printed circuits which is likely to be easily applied to SMA.

A further alternative process, although not based on solvent systems but which gives a very thin inert conformal coating, is the deposition of a polymer from monomer vapour under vacuum. The coating, which is condensed uniformly over all surfaces, is in the order of 5 microns thick. The material poly (para-xylylene) is known as Parylene and the product and process are available from Union Carbide Corporation. This approach, because it employs a vapour under vacuum, is more likely to penetrate the smaller gaps and so ensure all-round coverage. This can be particularly important with the small dimensioned spacings between SMD and the substrate surface[13].

8.5.1.4 MATERIALS FOR CONFORMAL COATING

It is not intended in this book to give details or selection aids to the merits of different types of coating material. The user must define the operating limits and assess the properties of each chemical type of coating. The principal properties are electrical, thermal and resistance to humidity, but factors such as ease of application, pot life, drying and cure temperatures, reworkability, resistance to fungus, and chemicals require to be considered. The range of types of coating and the physical properties within one type call for informed selection and it is fortunate that many surface coating suppliers have a comprehensive range and can give advice on selection. Basic types—acrylic, polyurethane, epoxide, and phenolic resins can often be obtained from within one surface coating supplier's range, while silicone and polyimide resins and Parylene are offered by specialised manufacturers.

8.5.2 Solids-based Conformal Coating

Dipping and fluidised bed methods for the application of materials, usually based on epoxide resin formulations, produce smooth, thick coatings which give chemical and mechanical protection to a circuit. Both methods require high temperature cures (125–180°C) and careful determination of a cure schedule to prevent 'blow-out' of entrapped air. The technique has been limited to small circuit boards and alumina substrates.

Thixotropic dipping formulations which exhibit non-slump at cure temperatures are two part systems and consequently have a limited pot life. The technique is suitable for both hand and automatic application to circuits.

Fluidised bed material is a powdered one component formulated system with a particle size typically 100–30 microns, which is fluidised by compressed air. The circuit is pre-heated and then dipped into the fluidised bed, which has been momentarily switched off to prevent contamination of the leadouts, and the assembly is then reheated to fuse the powder coating. The process is repeated until the required thickness of coating is obtained. Problems encountered are poor coverage due to shadowing by large components and differential coverage associated with the thermal capacities of different components.

8.5.3　Encapsulation

To produce a robust compact module, encapsulation in a resinous material has many advantages. This method can be adopted when coverage of a circuit by a simple box or case does not offer the degree of chemical and mechanical protection deemed necessary. Whether a re-usable mould process or an individual box is used to provide the external dimensions to the encapsulated module, a number of different types of liquid resin systems can be used. Two component systems of silicone, epoxide, polyurethane and polyester resins with the addition of a variety of inert mineral fillers offer a range of possibilities from which can be selected the optimum solution for the individual application. SMDs are not likely to demand any extra consideration compared with conventional assemblies.

Transfer moulding methods with materials formulated for the integrated circuit field will give full protection to surface mounted circuits. The bulk of the material of the encapsulant was thought to provide protection against humidity, the ingress of water being along the lead-outs. More recent studies with integrated circuits have questioned this assumption, but the level of moisture likely to penetrate through the bulk material is unlikely to affect encapsulated SMDs.

Some components are sensitive to the stress generated by the shrinkage of the resin encapsulant or the transfer moulding pressure. Glass diodes and ferrites are vulnerable to these effects, but localised pre-coating with a soft silicone rubber will reduce the stress concentrations as also will the use of softer encapsulant materials combined with the smaller dimensioned SMDs.

REFERENCES

1　'Automatic PC Board Test Equipment Speeds Production, Improves Quality', *AEU*, October (1983).
2　'Inspection Comes to the Surface', MME, *Electronics Manufacture & Test*, January (1985).
3　Morrisey, J. and Cobb, J., 'Automatic Visual Testing Equipment Designed for Loaded PCB Inspection', *Electri·onics*, p. 17, March (1985).
4　Klein Wassink, R. J., 'Soldering in Electronics', Electrochemical Publications, Ayr, Scotland, p. 424.
5　'Testability Critical for SMDs', *Computer Design*, p. 39, 15 July (1985).
6　Costello, J., 'Easy Access Aids the Testing of SMDs', *New Electronics*, November (1984).
7　Lawrence, B., 'Fixing the Future', *Electronics Manufacture & Test*, p. 66, February (1986).
8　Smith, M. and Cook, S., 'Probe Problems with SMDs', *Electronic Production*, p. 38, August (1985).
9　Solder-Wick Fine-Braid Solder Removal Co., Covina, CA, USA.
10　'Refining SMD Board Reworking', *Electronics Manufacture & Test*, p. 37, December (1985).
11　Rizzo, J. A., 'Repairability of Surface Mount Assemblies', Circuit Expo '84 Proceedings, p. 90.
12　Wargold, I., 'How to Use Conformal Coatings Efficiently', 'Printed Circuit Handbook', 2nd Edition, McGraw-Hill.
13　Cosens, M. E., 'Parylene Conformal Coatings', *Electronic Production*, p. 12, March (1983).

Chapter 9

THE PRACTICAL ADVANTAGES OF SMA

J. F. PAWLING and G. A. WILLARD

Mullard Mitcham, Mitcham, Surrey, England

9.1 OVERVIEW

When considering the advantages of SMA, a comparison with alternative fabrication technologies is implied. It is necessary to define a reference—is it the more conventional wire-ended components inserted into a printed circuit board, or a comparison against thin or thick film hybrids? Indeed, is it a question of 100% surface-mounted components or a mixed technology of SMD and wire-ended components, and what is the substrate—alumina, epoxy laminate, or metal? If the surface mountable device is first compared with its wire-ended counterpart, then an obvious advantage is size. The very fact of excluding the wire terminations provides a reduction in the area of substrate required. Earlier chapters (1, 3 and 4) also showed that generally SM Devices have smaller body dimensions as well. Thus, many of the advantages of SMA follow from this fundamental one of smaller components, and hence reduced substrate area. This is not true, however, when making a comparison with Thin and Thick Film Hybrids, when the components can be just as small and indeed may be the same ones! The advantage now becomes that of being able to use economic substrate materials with multiple conductor layers. As one uses smaller components and hence achieves denser packing, the interconnect problems become greater. These are resolved by resorting to several layers of interconnection and this currently is best and most economically executed through using epoxy laminated boards with plated-through holes.

Any comparison between SMA and alternative assembly technologies will vary depending on the complexity of the circuit, the need for ultimate miniaturisation and several other criteria, including reliability. The following paragraphs illustrate the advantages of SMA compared with alternative technologies using a number of practical examples. These are then summarised in a table.

9.2 MINIATURISATION

A pair of printed circuit boards are shown in Figure 9.1 using exactly the same electrical circuit. Both boards have conductor patterns on either side which are interconnected by plated-through holes. The smaller board achieves an area reduction of $3 \cdot 1$ times and a weight reduction of $3 \cdot 5$ times by the use of the smaller surface mountable devices wherever possible and also using both sides of the substrate for mounting the components. Chip resistors and multilayer capacitors are assembled on the bottom side of the epoxy board. Six Surface-Mountable SO packaged integrated circuits are mounted on the top side as are two DIL integrated circuits and eleven sundry components. The SO packages are first soldered onto the top surface and all the remaining components are then flow soldered from the opposite side. The number of components on both these circuit boards is exactly the same with the one exception that, while two separate 4K RAM (Random Access Memory) integrated circuits are used on the larger board, one 8K RAM has been used on the smaller one.

Here the obvious advantage of SMA is miniaturisation. The circuit is for a

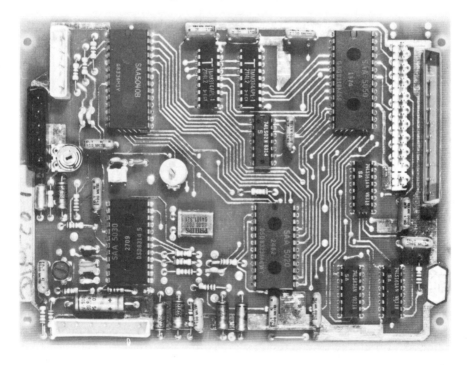

Fig. 9.1 (a) Two executions of the same electrical circuit. SMA has enabled a board area reduction of more than three times to be achieved. Here the top side of each board is shown. The solder sides are shown on the opposite page. (Courtesy of Mullard Ltd)

teletext decoder used in television receivers.[1] The smaller size offers greater freedom to the TV set designer, and the assembly is now sufficiently small and light to be plugged into the main mother board of the TV set by an edge connector, and it also opens up the possibility of providing teletext decoders in the smaller screen size sets. The manufacturing cost is also reduced; as less substrate is used, there is a consequential faster rate through the production processes, such as flow soldering, and assembly labour costs are less because of the enhanced speed of placement of the SM Devices.

This circuit is essentially a simple one. As the system becomes more complex the problem of interconnecting all the components while maintaining a high packing density becomes greater. Such a circuit, which is one used in a mobile

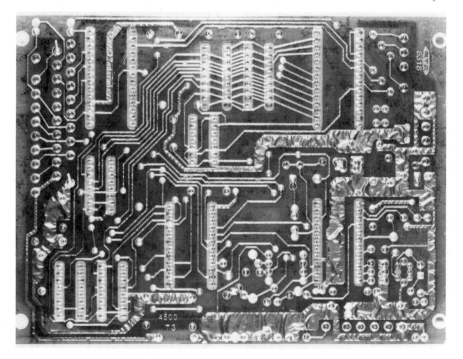

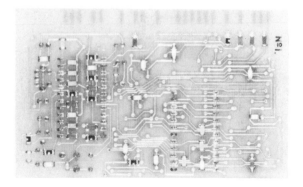

Fig. 9.1 (b) The solder side of the two executions opposite.

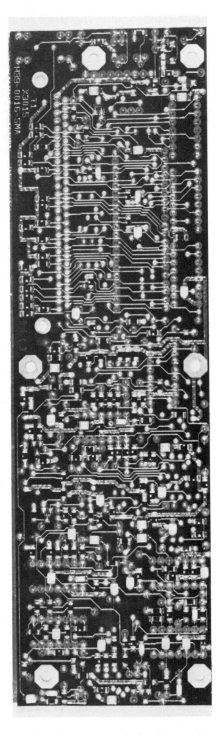

Fig. 9.2   A mobile radio board using SMA. 255 components are interconnected using four conductor layers on an area of 21 square inches. (Courtesy of Mullard Ltd)

radio, is shown in Figure 9.2. The number of components on this board has risen from the 60 used on the teletext assembly to 255 (4·25 times). The board area made available for component attachment is only 2·1 times greater so that the resulting component density has increased from just over 6 per square inch to over 12 per square inch. To achieve a satisfactory interconnection, four conductor layers are now necessary. The result is an electrically successful sub-assembly which is small and yet economic to produce. A further example of the application of SMA, to achieve a degree of miniaturisation where space is limited, is shown in Figure 9.3. In this case it is a module which can be mounted inside a domestic electricity meter case and is able to compute, store and display kW or kVA maximum demand.

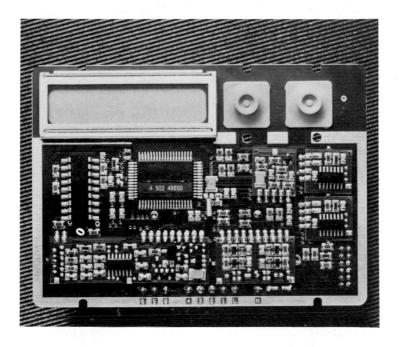

Fig. 9.3 A module board used inside an electricity meter cover employing SMA to achieve the small size required. (Courtesy of Landis & Gyr Ltd)

9.3 GREATER DESIGN FREEDOM

In Chapter 5, reference was made to the greater design freedom offered by SMA in being able to put components on either side of a substrate. With a mixed print approach, the designer can additionally choose between using a surface mounted chip component or a wire-ended device. Figure 9.4(a) compares the top side of two versions of exactly the same electrical circuit. In this example the conventional execution, the upper one, was designed first. (It was a requirement when subsequently executed in SMA mixed print that some parts of the circuit were not changed and that all external connectors were identical and in the same position.) Figure 9.4(b) on the following page compares the solder side of each board. To achieve the electrical interconnects with conventional components, a double layer of conductors was necessary. When later executing it in a mixed

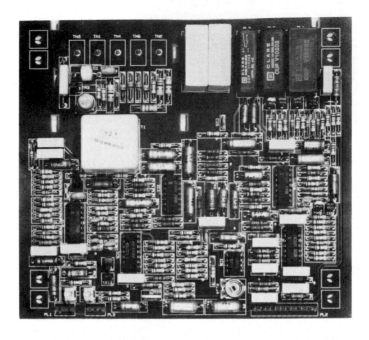

Fig. 9.4 (a) Two executions of the same electrical circuit. SMA has enabled a single conductor layer board to be used in place of a two layer board previously necessary. (Courtesy of Mullard Ltd)

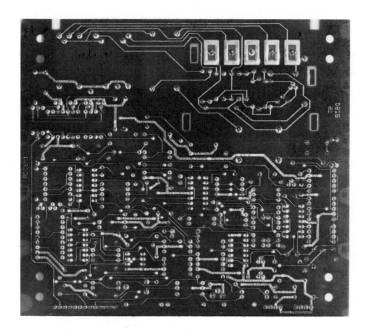

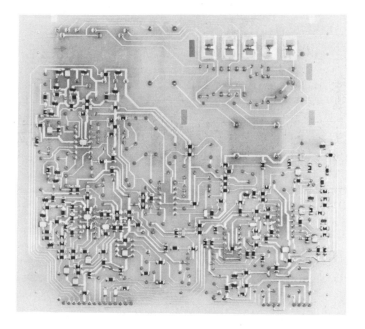

Fig. 9.4   (b) The solder side of the two executions.

print, the same layout designer was able to use a single layer conductor board. The reduction in cost of the printed circuit board by eliminating the need for a second layer of interconnection and the consequential plated-through holes was considerable. This was directly achievable because of the freedom of choice referred to above. In effect the equivalent of 1½ layers of interconnect were on offer, and in this particular application were sufficient. The main advantage in this application, a videotext telephone line coupling unit[2], was one of cost reduction.

9.4 REDUCTION IN COMPLEXITY

An SMA mixed technology sub-assembly is shown in Figure 9.5 consisting of 330 surface-mountable components, plus 68 DIL packaged ICs and over a 100 other components. SMA has been used to enable components to be mounted on both sides of the printed circuit board and so compact the circuit down to one substrate of 115 square inches. Without the use of SMA this area would have been considerably greater—too large for one circuit board. In fact, because of equipment space limitations, the circuit would have been executed on three separate boards with consequential electrical interconnections required between them. These connectors further increased the substrate area required, the cost of manufacture and chances of unreliable operation. So in this type of application SMA simplifies matters—fewer boards, fewer interconnections, less to go wrong and less to fault find in the event of failure. On the other hand, each board or module will be more expensive, having more circuitry on it and therefore fault finding in the field by board substitution methods will be correspondingly more expensive, when considering material costs alone.

9.5 COMPARISON WITH THIN AND THICK FILM HYBRIDS

Thin and Thick Film hybrid circuits use surface mounted components and hence are themselves surface mounted assemblies. They are a special case because the majority of resistors are inherently produced by a similar process to that used to generate the conductive interconnect and they are also constrained to use a flat, hard substrate material. The range of non surface mountable devices then becomes difficult to use as holes cannot be cheaply made in the substrate in order to mount the wire and tag-ended components. The manufacture of conducting and insulating layers is carried out sequentially and thus multiple interleaving of insulator and conductor layers becomes increasingly expensive. Any comparison is forced to the following conclusions. At one end of the spectrum of possibility there are resistor intensive circuits which are able to be interconnected in a single layer of conductive interconnect—and by definition these are not SMA and are best accomplished using Thin or Thick Film processes. At the opposite end of the spectrum are complex circuits involving hundreds of components, densely packed and requiring 4, 6 or more layers of conductors to interconnect the components electrically. These are best accomplished using surface-mounted components on multilayer printed circuit board substrates and, where necessary, wire or tag-ended components. In between these two extremes, there are a multitude of alternatives. Figure 9.6 shows an example of a simple circuit executed using both Thick Film and a mixed technology using surface mounted and wire-ended components on a printed circuit board, an example somewhere between the two ends of the postulated spectrum. In either execution, only one layer of interconnect is necessary. Thick film offers intrinsic electrical insulation on one

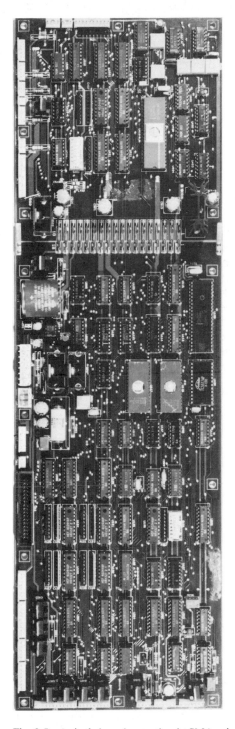 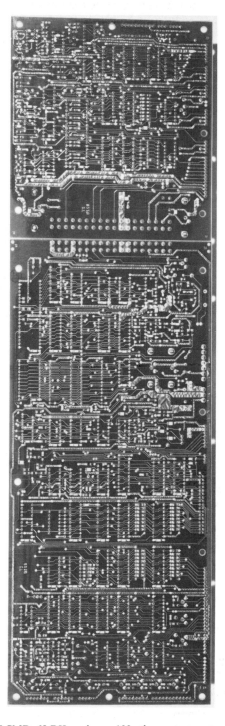

Fig. 9.5 A single board execution in SMA using 330 SMD, 68 DIL and over 100 other components. (Courtesy of Mullard Ltd)

SMA version
— Top side

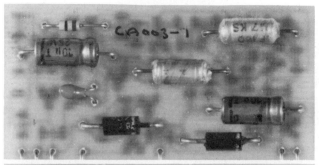

SMA version
— Solder side

Thick film
hybrid version

Fig. 9.6 Two executions of the same electrical circuit. One in SMA on PCB, one as a thick film hybrid. (Courtesy of Mullard Ltd)

side because of the alumina substrate. On the other hand, the non surface mountable components appear to be better assembled with good engineering principles in the PCB case. Both executions are of equal size but the PCB version can be expected to be cheaper to manufacture in production and to require lower initial tooling costs.

This section has limited itself to comparing SMA and hybrids as competing assembly technologies. It is additionally possible to consider hybrids as alternatives to silicon integrated circuits, particularly for an interim solution or in a low cost start-up situation.

9.6   UNUSUAL SHAPES

Most substrate materials by their intrinsic rigidity define a two-dimensional electronic assembly. Where space is limited, or is offered in an unusual shape, it can be desirable to curve round or even fold up a circuit. This requires a flexible substrate material such as polyimide or thin fibreglass. Before SMA, limitations existed in that wire terminated components presented severe problems to designers who wanted to bend their circuits. Indeed it usually meant that all components had to lie parallel to each other and with their axes perpendicular to the direction of folding. The small body size of surface-mounted components helps to remove this constraint and they can now be placed in either direction enabling circuits to be curled round much more easily. Figure 9.7 shows an

Fig. 9.7 An encapsulated flexible SMA formed to complete ⅞ths of the circumference of a circle of 35 mm diameter. (Courtesy of Mullard Ltd)

example of this type of circuit. It is, in fact, the layout shown in Chapter 1, Figure 1.10, but now wound to form ⅞ths of the circumference of a circle of 35 mm diameter. The unencapsulated assembly is again shown but this time viewed from the surface mount component side.

9.7 'SECOND THOUGHTS' DESIGNS

It is often the case, even if not with the reader of this book, that the final performance specification of a circuit is more complex and involved than originally conceived. Unfortunately, space is often allocated at an early stage in a development and, although sufficient space could have been allocated, designers suddenly seem to find too much electronics fighting for the too little space! The usual solution is to mount sub or daughter modules piggy-back fashion onto the main, or mother, substrate. With SMA, two possibilities occur, namely, make the daughter modules smaller, lighter and more easily mounted, or eliminate the need for the daughter board by mounting all the additional components on the main board. Figure 9.8 shows an example of an electronic programming module for a domestic washing machine—on the left the first production version using conventional components and a daughter board, and on the right a mixed technology version in which the sub-module has been eliminated offering potential advantages in cost and reliability.

Fig. 9.8 Two executions of the same electrical circuit. The use of SMA has eliminated the need for the daughter board. (Courtesy of Mullard Ltd)

9.8 TABULAR COMPARISON

Table 9.1 attempts to summarise the points made with the examples already described. It uses only comparative terms; the final choice of technology will be dependent on many parameters, too many to consider within the limitations of being expressed in such a simple table.

9.9 COST COMPARISONS

Comparisons of cost, albeit rather crude, of SMA with other assembly technologies are included in Table 9.1. Different companies will calculate costs in different ways, reflecting their varying policies on capital depreciation, subsidising new products with established ones, using government and other third party financing and so on.

Table 9.1

	Wire-ended Components + PCB	Surface Mounted Components + Thin Film	Surface Mounted Component + Thick Film	Surface Mounted Components + PCB	SMC + Wire-ended Components + PCB (Mixed Technology)
Simple circuits— 1 inter-connection layer	Good	Good	Good	Good	Good
Medium complex circuits—2 inter-connection layers	Excellent	Difficult	Difficult	Excellent	Excellent
Complex circuits— 4 or more inter-connection layers	Excellent	Increasingly difficult	Increasingly difficult	Excellent	Excellent
Miniaturisation	Difficult	Excellent	Excellent	Excellent	Good
Unit costs in large-scale production	Low	High	Medium	Low	Lowest
Tooling costs	Low	Medium	Medium	Low	Low
Process machine investment costs	Can be very low	High	Medium-high	Medium	Medium
Layout design know-how requirements	Low	Medium	Medium	Low-Medium	Low-Medium

The cost of manufacture of any electronic assembly will include the following:

Material costs—bought-in 'off the shelf' components, e.g., transistors, passives, ICs
—bought-in custom designed components, e.g., PCBs
—indirect processing materials, e.g., solder
Direct labour costs of assembly, test, repair, packing etc.
Indirect labour costs of management, buying, selling, accountancy, etc.
Depreciation costs of machines bought or hired to build the product.
General overheads of rent, taxes, heating, lighting, etc.
Cost of designing the product.

The above list excludes selling and other expenses, covering only the items involved in design and manufacture. This discussion will limit considerations to four of the above, namely Bought-in Materials, Direct Labour, Depreciation on Machines and Design Costs.

These are the most fundamental, generally are the largest, and can vary significantly with the chosen technology. It is very difficult and many would say foolish to attempt to give typical values for the above, but to give the reader some feel it *could* be that the cost of production of an electronic sub-assembly in large scale production (100,000 or more per annum) would be split as shown in Figure 9.9.

The four elements will now be considered in turn and the likely price trends and reasons for these will be estimated.

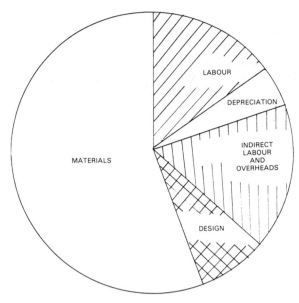

Fig. 9.9 Possible manufacturing cost breakdown of an SMA in large-scale production.

9.9.1 Materials

The purchased materials are primarily substrates, passive components, discrete semiconductors and integrated circuits. Encapsulated semiconductors such as SOT23 transistors and diodes, and SO integrated circuits are just smaller versions of the earlier tag-ended devices. Using less material they are fundamentally cheaper[3]. MELF resistors are wire-ended resistors without wires—again a case for reduced component prices. It is often difficult to compare a 68- or 84-pin chip carrier or flat-pack IC against a DIL equivalent, as the latter may not exist, but perhaps a satisfactory comparison is between one multi-pin surface mountable IC encapsulation against the use of several 40 and 28 pin DILs. In this case the single encapsulation solution will be cheaper.

The substrate will always be smaller with SMA than with wire-ended components. Less material will again mean a lower price. Thus the material cost of an SMA should be lower than alternative technology approaches. The only restriction is in the interim introduction of new techniques and components when the quantities being manufactured are lower than long established methods and the cost benefits of large scale production have not yet been achieved.

9.9.2 Direct Labour

The labour content of an electronic sub-assembly is split between the actual putting together and the testing to ensure that the product has been assembled correctly. The multi-headed SMA component placement machines described in Chapter 6 have component assembly rates exceeding 500,000 per hour. These are well in excess of those achieved in inserting wire-ended components, and so the labour content can be much reduced. As an example, a single-headed axial component insertion machine can be expected to insert a component on average in 0·6 of a second, whereas a surface mount equivalent can place a component every 0·08 of a second.

The processes that follow component assembly in a production line will be very similar for both SMA and conventional printed circuits, i.e., flow or reflow soldering, cleaning, in-circuit component testing and functional testing. In practice, further cost savings will ensue with SMA. If an SMA version is half the size of a more conventional assembly, then the same production belt can process twice as many devices in a given time, or alternatively the belt can be narrower, and a cheaper machine employed.

9.9.3 Capital Depreciation

Trends in capital depreciation costs are more difficult to analyse. SMA demands automatic assembly machines whereas hand labour can still be employed for the insertion of wire and tag-ended components. The minimum capital investment costs are therefore greater but the consequential reduction in labour costs will more than offset the investment price provided throughput quantities are high enough. This will be particularly so if double or treble shift working can be arranged. Also it should be remembered that one SMD placement machine, if chosen wisely, will place a large range of different types of component. With conventional components they are so different in shape and size that a production line may require one machine for axial components, another for radials, a third for ICs and so on. As mentioned in the previous section, savings may be possible in the area of soldering and cleaning because of the reduced substrate area of the products. Smaller or fewer machines may be needed. With testing there is likely to be little difference as the same parameters need to be checked with any specific product. In-circuit computer-aided testing can be more expensive, as double-sided test jigs, or two passes through the tester, will be necessary if components are mounted on both sides of the substrate. As a guide to costs one reference[4] has put the cost of equipping a prototype SMA laboratory at $200,000 and a medium-volume manufacturing facility at nearly three quarters of a million US dollars.

9.9.4 Design Costs

The system design costs are independent of the electronic sub-assembly's technological execution, i.e., the generation of an agreed specification, choice of circuit design, simulation and verification. The physical layout design will take longer with SMA if the ultimate benefits of miniaturisation are attempted. More powerful computer-aided design machines will also be required. This part of the design is likely to be more expensive, but on the other hand is generally not a significant part of the total design cost.

Initial hardware samples cannot be easily produced by purely manual methods. It is generally cheaper and quicker to programme a software controlled chip placement machine, even for 6 to 10 samples. These are additional design costs when using SMA compared with more conventional methods, but they will need to be spent at a later stage in any case. The largest investment is in acquiring the technology 'know how' to make surface mounted assemblies with high efficiency. Generally, these costs are recovered over a number of products rather than the first one to use SMA.

9.10 RELIABILITY

The reliability of a Surface Mounted Electronic Assembly is a function of the intrinsic reliability of the components used and the reliability of the attachment processes. It is, of course, theoretically possible for the process of attachment to have a detrimental effect on the reliability of the components. For example, too long immersion in solder at too high a temperature can affect semiconductors or plastic film capacitors. Reference is made in Chapters 3 and 4 regarding the reliability figures for the Surface Mounted Components themselves. The simple passive components such as chip resistors and capacitors achieve Typical Failure Rates of $<1 \times 10^{-9}$/component hour[5], while plastic encapsulated integrated circuits have typical figures of $<5 \times 10^{-6}$/component hour[6]. The main difference in the reliability of an SMA version of an electronic circuit compared with alternative technologies will be due to the attachment process and the matching of component and substrate to which it is attached. This is a complex issue, introduced in Chapter 2.1 and discussed further in this chapter, and depends upon whether, for example, one is considering a small component (chip resistor), or a large one (68-pin chip carrier); a component with or without flexible leads, and whether the substrate is rigid or flexible. First of all, how does one check or measure the reliability of the attachment process?

9.10.1 Testing for Reliability

There can be four main experimental methods of assessing the reliability:

Powered Functional Cycling
Temperature Cycling
Mechanical Cycling
Step Stress Testing

Powered Functional Cycling is the nearest to actual life that the equipment can experience. Because it is close to the 'real thing' it is also time consuming as, unless there is a fundamental weakness, the equivalent of years of real life may be necessary to obtain failures.

Temperature cycling in a laboratory test chamber is quick in comparison and requires less effort. The danger is that, if the maximum and minimum temperatures are too unrepresentative of actual experience, non-representative failures may be obtained.

Mechanical flexing and vibration is the quickest test of all, but can be even more non-representative of normal equipment existence.

Step Stress Testing is a technique used to induce failures by overstressing the product. Investigation of these failures, and subsequent modification of the design can be used to produce a more reliable one. As this type of testing is designed to produce failures, publication of the results can be open to misinterpretation. It nevertheless is a powerful tool to enable designers to achieve a very reliable product and has been used in applications as diverse as consumer electronics and space exploration.

Returning to temperature and mechanical cycling, then a typical test sequence (based[7] on IEC68-2) used to evaluate SMA would be as follows:

1 Temperature Cycling
   $-20°C$ to $+70°$ with 1 hour dwell time. 1000 cycles

2 Thermal Shock
   (i) $-20°C$ to $+70°C$. Immediate Transfer. 10 cycles.
   (ii) $-20°C$ to $+125°C$. Immediate Transfer. 5 cycles.

3 Vibration
   5 Hz to 150 Hz and back to 5 Hz at $0·75$ mm displacement, constant acceleration of 10 g, frequency sweep 1 octave/minute 36 cycles.

4 Bump
   40 g. 4,000 Bumps (split equally between each of the 6 directions in 3 mutually perpendicular planes).

Additional tests could be humidity (alternating high humidity such as 95% relative humidity at high temperature with lower temperature conditions to encourage condensation during the cold cycle), powered life tests, and of course visual inspection including the sectioning of solder joints.

9.10.2   What Problems to Expect

The range of substrate materials can be many and various and can include:

Rigid epoxy fibreglass
Rigid polyester fibreglass
Rigid paper-phenolic
Flexible polyimide
Flexible epoxy fibreglass
Alumina
Beryllia
Various coated metals
Injection moulded plastics.

The problems of successfully attaching components to such a range of materials will depend on the material selected, particularly upon its thermal coefficient of expansion (Tce), and whether it is rigid or flexible and compliant. This is because with thermal cycling, either due to variations of the ambient temperature, or through internal heating of the component from electrical power dissipation, unless component and substrate have near identical Tce, thermal mismatch will cause eventual failure of the assembly.

This thermal mismatch problem is particularly significant when the component is large and directly soldered to the substrate. An enormous amount has been written on the problem of attaching ceramic chip carriers onto Printing Wiring Boards. The carrier[8] has a Tce of 6 to 7 ppm/°C, the silicon chip inside has a Tce of about $2·7$ and an epoxy wiring board $14-15$. There has to be a mismatch somewhere! One answer is to use ceramic chip carriers only with substrates having similar expansion coefficients such as ceramic or copper clad Invar. With this latter substrate, copper clad Invar sheet (65% iron, 35% nickel alloy) is sandwiched between epoxy-glass or polyimide-glass laminates. By varying the proportions of copper and Invar the Tce can be adjusted to optimally match that of the chip carrier.

The solutions with epoxy fibreglass boards are to use an IC encapsulation with leads which will absorb the thermal mismatch stress (e.g., leaded plastic chip carrier or SO), or to bond a 50 micron (0·002 in.) thick compliant elastomeric layer to the epoxy glass.[8] Electrically conducting materials such as carbon or metal are introduced into the elastomer where necessary for electrical contact. A third but only partial solution is to increase the height of the solder joint. It has been predicted that, for a 0·85 in. square ceramic chip carrier mounted directly onto epoxy glass, the number of cycles before failure can increase from 100 with 0·004 inch solder joint height to 1000 cycles when the solder joint is 0·01 in. high.[9]

The afore-mentioned thermal mismatch problems recede as the size of the component becomes smaller. Chip resistors, SOT23 transistors and small multilayer capacitors should offer no problems even on epoxy glass. The Mullard organisation subjected a range of such small components mounted on epoxy and phenolic printed wiring boards to over 4 million solder joint thermal cycles. One joint failed and this was because the copper pad had poor solderability and the solder joint was always a poor one. This one failure underlines the basic rule in achieving high reliability with any technology: process conditions must be strictly controlled and cleanliness maintained at all times.

It is generally accepted that high reject rates in manufacture will lead to poor long term reliability for those products tested initially good, and *vice versa*. Products which can be made with high efficiency will usually have a long and healthy life. Good solder connections are dependent on accurate component placement, adequate control of the adhesive application (when used with wave soldering) as well as good surface conditions for wetting and close control of the soldering machines. Placement reliabilities of 10-20 ppm are already claimed by Philips for their assembly machines[10], and it should therefore be possible to target for an overall figure for connection failure of 50 ppm—a figure that may seem good until it is realised that there can be 2 or 3 thousand connections on an SMA and thus one in every seven or so boards would still be a failure.

The greater miniaturisation possibilities achievable with SMA enable one hundred per cent Component Burn-in to become more economic as less oven space is required for a given number of assemblies. Burn-in is based upon the theory that the rate of physico-chemical reactions which are the cause of infant mortality in electronic assemblies is increased by operating the product electrically while maintaining a high ambient temperature[11], e.g., 8 months at 25°C is equivalent to one week at 150°C. Thus, early so-called 'bath tub' field failures can be screened out in the factory prior to delivery, providing sufficient oven space is made available for one or more week's production.

9.10.3 Thermo-mechanical Reliability Test Results

It has been stated that Surface Mounted components should have similar, if not superior, performances to their DIP counterparts in most applications.[12] Information published in various articles and reports[13] backs up this claim and a series of practical tests are now described to give further weight to this statement.

A number of Mullard Teletext Decoder Modules (Figure 9.10) made using the Mixprint assembly technique were subjected to a combination of thermal and mechanical cycle tests by an independent test laboratory[14] in order to assess their performance. The tests and the results obtained are now described, giving practical figures to an actual SMA situation.

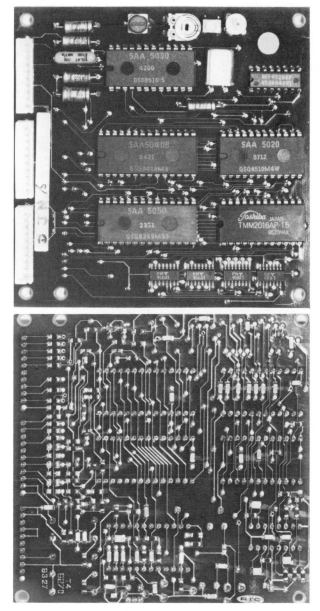

Fig. 9.10 Component side and solder (under) side of Mixprint Teletext Decoder board (Courtesy of Mullard Ltd)

The boards were subjected to the following cycling conditions:

(i) Thermal cycling
    −20°C to +70°C with a 9 minute ramp and 10 minute dwell time, total cycle 38 minutes.

(ii) Mechanical flexure
    (a) 55 cm radius at a frequency of 0·5 Hz

(b)   55 cm radius at a frequency of 7·5 Hz
(c) 202 cm radius at a frequency of 0·5 Hz
(d) 202 cm radius at a frequency of 7·5 Hz
  (55 cm radius = 2·75 mm flexure over board length of 110 mm)
  (202 cm radius = 0·75 mm flexure over board length of 110 mm)

A total of 9 boards were subjected to combinations of these cycling tests as shown in the following table.

Table 9.2

No. of Cycles (Test Sequences)	Thermal	Flex (a)	Flex (b)	Flex (c)	Flex (d)
Board 1	1600				
2		12000			
3			12000		
4	100 (1)	1600 (2)			
5	100 (1)		1600 (2)		
6				12000	
7					12000
8	100 (1)			1600 (2)	

(The quantity of test cycles is shown followed by a figure in brackets giving the sequence of testing.)

The boards were visually inspected after every 200 cycles and the positions of any cracks in the solder joints noted. Twelve chip resistors and capacitors were monitored electrically during the testing to check for any variation.

9.10.3.1   THERMAL CYCLING RESULTS

Boards 1, 4, 5 and 8 were all thermally cycled while 4, 5 and 8 were also flexure cycled.

Board 1, which was thermally cycled, may be taken as a benchmark for the other 3 boards which also received flexure cycling. The visually observed cracking rates are shown in Figure 9.11. Cracks were first observed at 900 thermal cycles, after which they developed quite rapidly in the solder joints affected. The joints that cracked most severely were, significantly, not surface mounted joints but through-hole joints of the DIL packages. These only developed seriously on the solder side (lower side) of the board where most of the SM solder joints were also found. The joints of the 1206 capacitors and resistors also manifested cracking to a lesser degree while none of the SOT23 or SOIC joints showed any tendency to crack under thermal cycling. Although the DIL cracking seemed to level off at around 75%, 4 out of 6 of the DIL devices had shown 100% cracking in all their leads (lower side of PTH solder joint).

Board 4, which in addition to thermal cycling received a similar number of low frequency (0·5 Hz), high amplitude (2·75 mm) flex cycles, did not differ very significantly in result from Board 1, with the exception that significantly more cracking occurred on the component side of the DIL devices (Figure 9.12). There

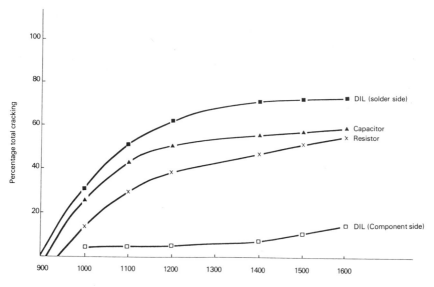

Fig. 9.11 Visually observed cracking rate of components on Board 1 caused by thermal cycling from −20°C to +70°C. (Courtesy of Plessey Research (Caswell) Ltd)

was also an isolated crack in one of the SOIC leads, and 100% cracking in 4 of the DILs was again observed.

Boards 5 and 8, which in addition to thermal cycling received low amplitude flexure cycles, were each very similar in result to Board 1. In fact, the additional flexure in these boards did not appear to affect the cracking of component joints significantly during thermal cycling at all.

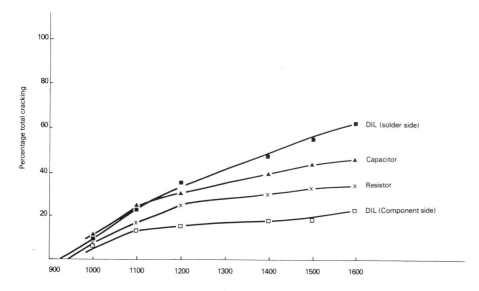

Fig. 9.12 Visually observed cracking of Board 4 caused by a combination of thermal cycling and low frequency flexure cycles. (Courtesy of Plessey Research (Caswell) Ltd)

9.10.3.2   FLEXURE CYCLING RESULTS

A very much larger number of flexure cycles than thermal cycles were required before the onset of solder joint cracking was observed in these boards. Flexure cycling can, however, by comparison with thermal cycling, be achieved relatively rapidly and therefore large numbers of cycles could, in a practical application, be completed over a relatively short period.

Board 2 was subjected to the low frequency (0·5 Hz) high amplitude (2·75 mm) flexure cycle which, on the basis of previous work, would be expected to be the most damaging of the flexure cycles. Earlier work has indicated that low frequencies tend to be more damaging in flexure than higher frequencies. The first cracking that was observed occurred at 6000 cycles. This was particularly marked on the top (component) side of DIL integrated circuits where considerable cracking was evident. This is in contrast to thermal cycling where DIL cracking occurred almost exclusively on the solder side of the board. In addition, the chip resistors and SOIC components also showed significant cracking by 12,000 cycles. By contrast the chip capacitors showed only minimal cracking by 12,000 cycles. Figure 9.13a shows the cracking development rates of the various components between 6000–12000 cycles.

Board 3, which was subjected to high frequency (7·5 Hz), high amplitude flexure cycling showed a similar but reduced crack development to board 2 with the exception that barely any crack development occurred in the SOIC leads. Therefore by 12,000 cycles 30% cracking had occurred on the Board 2 SOICs, whereas only one single crack (2%) had developed on the Board 3 SOIC lead solder joints. The crack development rates of Board 2 and Board 3 are shown in Figure 9.13(a) and (b). Boards 6 and 7 were subjected to low frequency, low amplitude and high frequency, low amplitude respectively. Very little crack development had occurred in either of these boards by 12,000 cycles. Some limited cracking had, however, occurred on the component side DIL joints; it was more severe in Board 6 which received the low frequency cycle.

9.10.3.3.   ELECTRICAL RESULTS

An array of 12 chip resistors and 12 chip capacitors orientated in opposing directions was monitored electrically during each test. Those boards which were subjected to mechanical flexure (seven out of the eight) were electrically probed while held in a fully positive flexed position. No significant variations in the values (±2%) of any of these chip components were recorded during the test.

9.10.3.4   INTERPRETATION OF THE RESULTS OBTAINED

In addition to the results recorded, it became clear for the boards which were subjected to high amplitude flexure that the cracking in the chip components occurred preferentially in the areas of board that were not located below DIL components. This effect is almost certainly due to the stiffening effect that the DIL imparts to the area of PCB immediately below it. In other words the area of board not covered by a DIL package curves the most during flexure. Figure 9.14 on page 222 illustrates this effect by showing the position of cracking relative to the areas shielded by the DILs. This would appear to be the significant factor in the observation that under flexure cycling the chip capacitors cracked significantly less than the chip resistors (see Figure 9.13). The fact that 72% of the

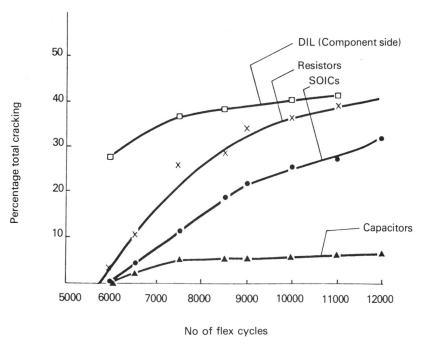

(a) Board 2 (Low frequency, high amplitude)

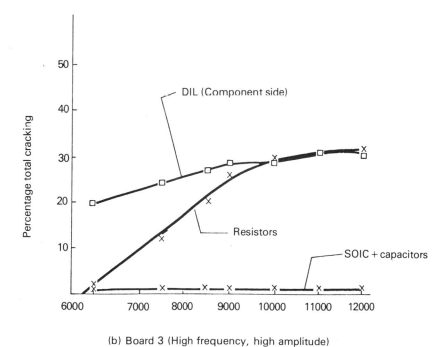

(b) Board 3 (High frequency, high amplitude)

Fig. 9.13 Comparison of crack development in Boards 2 and 3 due to mechanical flexure only. (Courtesy of Plessey Research (Caswell) Ltd)

222   *The Practical Advantages of SMA*

chip capacitors are located immediately below the DILs, whereas only 12% of the resistors are, means that a much larger proportion of the capacitors than resistors are protected by the DILs during flexure. Such protection is not, of course, imparted during thermal cycling and consequently the crack development rates under such conditions are very similar for chip capacitors and resistors (see Figure 9.11).

The observation that DIL component joints crack more severely than SM components during thermal cycling is interesting, perhaps illustrating that a cracked solder joint does not necessarily represent a reliability hazard, in view of the fact that DILs have operated with acceptable reliability in the field for a very considerable time. However, as long as cracks exist only in the solder fillet on one

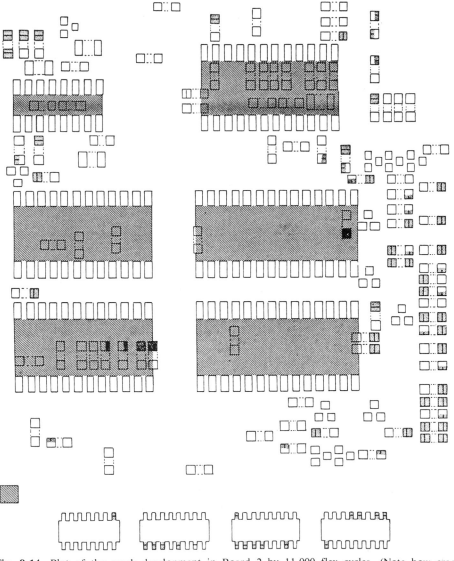

Fig. 9.14 Plot of the crack development in Board 2 by 11,000 flex cycles. (Note how crack development is chiefly in joints of components not protected by DILs) (Courtesy of Plessey Research (Caswell) Ltd)

side of the circuit board, and as long as the copper PTH board remains intact, it is clear there can be no electrical discontinuity.

SOIC component joints showed no tendency to fail during thermal cycling and the joints were still in excellent condition after 1600 cycles. Flexure cycling by comparison did damage the joints, although this was only really significant on the low frequency, high amplitude cycling. Figure 9.13 indicated how, by 12,000 cycles, over 30% of SOIC solder joints were cracked on Board 2, whereas less than 2% had cracked on Board 3. Figure 9.15 shows typical SO cracking on Board 2.

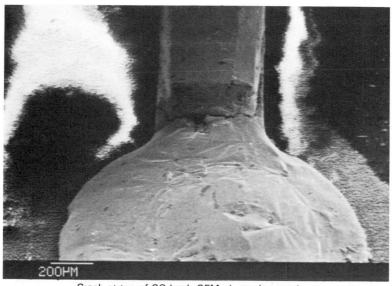

Crack at toe of SO lead. SEM photomicrograph.

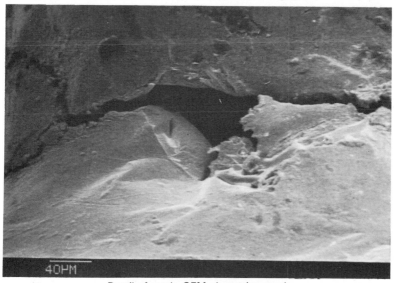

Detail of crack. SEM photomicrograph.

Fig. 9.15 Photomicrographs of SOIC joints cracked during flexure cycling. (Courtesy of Plessey Research (Caswell) Ltd)

Flexing has not been found to accelerate the effects of thermal cycling damage significantly. This is perhaps not surprising in view of the fact that flexing tends to be considerably less damaging than thermal cycling under the conditions employed in this test. The possibility that cracks initiated by thermal cycling could be propagated by flexing seems feasible, but has not been observed. The observation that a considerable number of cracks develop in both chip resistor and capacitor solder joints during thermal cycling or flexure cycling could be a considerable reliability concern for SM assemblies. In practice, the cracks

Cracked 1206 resistor joint. SEM micrograph.

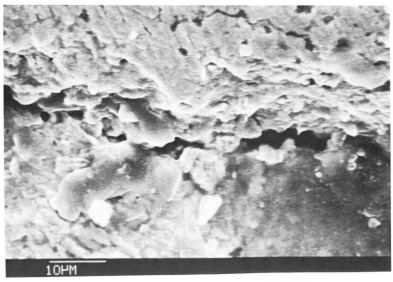

Detail of crack.

Fig. 9.16 Photomicrographs of chip resistor cracked during flexure cycling. (Courtesy of Plessey Research (Caswell) Ltd)

observed have not looked serious on the surface, having more the appearance of creases or surface disruptions on the solder joint, a typical example being shown in Figure 9.16. Further, after initiation, the cracks were rarely observed to become worse, almost as if the surface deformation in the solder acts as a strain relieving device. Metallographic sectioning of the joints has not revealed any internal cracks in the chip component solder joints at all, and no electrical failure has been observed either.

9.10.3.5 CONCLUSIONS DRAWN FROM THE PRACTICAL TESTS

Solder joint cracking is observed in both leaded (DIL) and surface mounted components on printed circuit boards as a result of both thermal cycling and flexure cycling. The surface mounted components showed less cracking than the DIL ICs in these tests.

Cracks developed in the joints of chip passive components during thermal cycling, but the slight flexibility imparted by the leads of SO devices meant that these components did not show any cracking.

Flexure cycling produced cracking in both passive components and SO IC leads, although a considerably larger number of cycles were required to initiate damage than with temperature cycling. Low frequency flexure (less than 1 Hz) is more damaging to solder joints than high frequency (greater than 5 Hz).

There was no evidence of component failure during the test. None of the devices monitored showed any significant change in value during the testing.

The Mixprint style of assembly used for this PCB assembly has advantages over a 100% SMA, in that the DIL ICs impart some protection against flexure damage to SMAs on the other side of the board beneath them, by effectively locally stiffening the PCB.

Further investigations by sectioning of the cracks formed in the test cycles confirmed that they were a surface phenomenon only.

It can therefore be concluded from these tests that Mixprint assemblies made incorporating SMDs on double-sided fibreglass boards perform adequately under the thermo-mechanical environments covered by the conducted tests. These boards were produced by automatic component pick and place machinery and wave soldering and these results apply to this type of technology. Extrapolation to other methods of assembly should be used as an indication only and is not really to be recommended because of the problems of correlation of conditions encountered.

9.10.4 Summary

It is essential, particularly if large temperature extremes are likely to be encountered, to select components and substrate material wisely. Provided this is done, then the automatic assembly and soldering inherent with SMA will ensure a consistently high quality product, and in turn a high reliability one.

REFERENCES

1 Development Sample Data 6101VM, Mullard, November (1979).
2 Development Sample Data VM6510, Mullard, November (1983).
3 'Surface Mount—The Key to the Future of the PCB Industry', *New Electronics*, p. 59, 7 January (1986).

4 'Choosing SMT and Setting Up a Facility', *Electronic Packaging & Production*, p. 74, January (1986).
5 'Surface Mounted Devices. Resistors', Philips, 9398 316 90011, May (1983).
6 'Surface Mounted Devices. Integrated Circuits', 9398 316 40011, May (1983).
7 'Basic Environmental Testing Procedure', JEC-68-2 or BS 2011, Part 2.
8 *Electronic Design*, p. 248, 10 January (1985).
9 'Surface Mount Technology', ISHM, Technical Monograph, p. 103.
10 'Modular MSD Placement Systems', Philips, 9398 216 70011, September (1983), Reprint June (1984).
11 Wood, E. R., 'Component Burn-in Using Fluorinert Liquids', Report 3M Commercial Chemicals.
12 'How to Use Surface Mount Technology', ISBN 0-904047-44-X, Texas Instruments, Dallas, Texas.
13 Reliability Report, Siliconix Inc., Santa Clara, California, May (1985).
14 Brierley, C. J., 'Thermo-Mechanical Reliability Testing Report', Plessey Research (Caswell) Ltd, Allen Clark Research Centre, Caswell, Towcester, Northants.

Chapter 10

A PRACTICAL APPROACH

J. F. PAWLING and G. A. WILLARD

Mullard Mitcham, Mitcham, Surrey, England

10.1 THE PRODUCT LIFE CYCLE

Any electronic equipment product which is designed, manufactured, and used, has a Product Life Cycle. Throughout the earlier steps of this cycle, Go/No-Go decision points occur when life can be allowed to continue or be aborted. Where the full life span occurs, then it will typically be as shown in Figure 10.1. The first

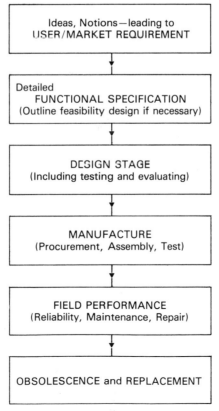

Fig. 10.1 Typical product life cycle.

stage involves ideas or notions which gradually harden to become a User or Market Requirement describing what is wanted and at what price. This User Requirement is reflected at the next level of development as a Detailed Specification. Ideally this is written by the product design team and agreed by the customer or his representative(s). The more detailed this specification is made, the less chance there is of the wrong product being designed. In order to arrive at a specification which the designers know can be met, they may undertake an outline feasibility design to check specific aspects which are particularly new or potentially difficult. When, and only when, a detailed specification has been fully understood and agreed, should the main design phase begin.

A modern electronic system is likely to encompass software, mechanical, circuit, and technological design aspects. As these areas are usually tackled by a team of people offering differing technical expertise, it is necessary to break down the design tasks into detailed and defined modules of work which can be tackled independently by different types of engineer. As the design work progresses, so each completed part needs to be separately tested and evaluated, and then tested again, as the various sections are integrated until the final complete product materialises as a prototype to be fully evaluated against the performance described in the Specification.

In parallel with the design phase a manufacturing capability needs to become available. This may already exist in terms of machinery and people and only require some specific jigs, fixtures and software programmes. At the other extreme, an entirely new set of process machines may need to be commissioned. The introduction of the new product in the manufacturing area will almost invariably require decisions on how it is to be tested.

Finally, the product is launched into the market place. Earlier work would have ensured, as much as possible, that a high reliability is guaranteed under the conditions it is going to experience. In practice, some field maintenance is usually necessary and at an early stage decisions will have been made, concerning how repairs will be carried out and who is sufficiently experienced to do such work. Maintenance may also imply minor design modifications on later models resulting from field experience or pressure for additional and improved features as a result of competitors' efforts. Finally, the product will become obsolescent and be replaced by a new design. This chapter will take a theoretical SMA product and consider each of the life cycle stages in turn, often referring to previous chapters in the book.

10.2   MARKET REQUIREMENT AND DETAILED SPECIFICATION

The User or Market Requirement should be a written document produced by the client or his representative/surrogate, using his language and methods of description. It may be necessary for the designer to interrogate the customer in order to ensure that he has thought sufficiently deeply about what he wants. This document is not the specification to which the design team will work. It will often be incomplete, have redundancy and be insufficiently exact. The Design Team, or its Project Leader, must now take this document and prepare a detailed specification. This new document must be exact and complete. It should be clear to anyone reading such a document what performance is required, what limits are specified with regard to size, weight, rigidity, flame retardancy, connectivity and so on. The environment to which the product will be subjected must be clearly stated together with the reliability performance expected. Finally, but not least, the required price and timescale targets must be clearly stated. An attractive product which is too costly or arrives too late to obtain the necessary market share can be as unacceptable as a product that performs inadequately.

10.3 REASONS FOR SELECTING SMA

During the writing of the Detailed Specification, it may be necessary for the designer to check the feasibility of being able to achieve the requested performance. It is at this stage that alternative technological solutions for executing electronic hardware will be evaluated. Chapter 9 considered some of the practical advantages of SMA. These included miniaturisation, greater design freedom (with a mix print approach), reduced complexity and the ease of handling unusual shapes. The designer can question whether the use of SMDs 100% or a mix of SMDs and leaded components will offer the best solution. He can choose his substrate material from a large range of possibilities (see Chapter 2), knowing that SMDs can be used on both rigid and flexible materials and that a minimum of substrate area will be utilised at the lowest cost.

Let us throughout this chapter take a specific example, for instance a portable, mains powered product, used in the field of micro computing. Size and weight need to be reduced to a minimum, yet cost must be low as the product is to be sold to domestic and small business users. Sales and hence production quantities should be high, and automatic assembly and test methods can be used to help in keeping the manufacturing costs low. Supposing the system designer has already indicated that his circuit will consist of 30 integrated circuits and 120 other components and is substantially a logic sub-system, the layout designer would then consider an SMA approach attractive as components could be mounted on both sides of the substrate in order to reduce the overall dimensions. The majority of components would be SMDs, yet the few that could not be, for various reasons, can be conveniently inserted by choosing an inexpensive fibreglass epoxy printed circuit board substrate with plated-through holes. These holes can be used for mounting the leaded components and for supplying vias between the two conductor patterns on either side of the board.

10.4 THE DESIGN STAGE

The design of the product will be concerned, in general, with all the relevant aspects of the device, including the Electrical performance and Mechanical construction. The requirement and constraints of these will normally define the substrate size and shape.

The example outlined in the previous section will be presented to the Assembly Designer in the form of a complete circuit diagram and a functional component parts list. Figure 10.2 shows a portion of a typical list. This functional component parts list must be converted into an actual parts list for the computer-aided substrate layout system to take as its data input. The component selection process must take account of any electrical constraints imposed by the circuit design, and mechanical restrictions due to the system requirements.

The example assembly, previously defined, with its 30 integrated circuits and 120 other components, would be evaluated in order to decide on the type of substrate required and the number of interconnection layers needed. The circuit has some leaded devices and hence needs a substrate which can be provided with holes for their leads. The need to keep the cost at a reasonable level means that a

INTEGRATED CIRCUIT	80C31 MICROPROCESSOR	IC3
" "	2764 ROM	IC4
" "	74LS257	IC6, IC7
" "	74LS573	IC8
" "	74LS08	IC5
TRANSISTOR	BC548	TR1, TR3, TR7
"	BC558	TR2, TR5, TR6, TR8
ZENER DIODE	1N4371	ZD1, ZD5
DIODE	1N914	D1, D2, D3, D4
CAPACITOR	100 μF 16 V ELECTROLYTIC	C8, C14, C18
"	0·1 μF 50 V CERAMIC	C1, C2, C3, C6, C10
"	10 pF 50 V CERAMIC	C12, C13
RESISTOR	1K5 0·25 W 5% METAL FILM	R10, R12, R14
"	10K 0·25 W 5% METAL FILM	R1, R2, R3, R4, R13
"	470K 0·25W 5% METAL FILM	R11, R18, R20
"	100R 0·5W 10% CARBON FILM	R19
CRYSTAL	10 MHz	XL1

Fig. 10.2 Part of a typical functional component parts list drawn up by the circuit design engineer to define the components needed to implement his circuit.

single-sided phenolic paper printed circuit board would be ideal, but the interconnection problems of 30 integrated circuits would almost certainly preclude the use of only one layer. A multilayer board would be too expensive for the application and is probably not necessary. The example would therefore be made as a Mixprint assembly using both Surface Mounted and Leaded components on a double-sided plated-through-hole fibre glass printed circuit board. The next paragraphs will now consider this Mixprint assembly in more detail.

10.4.1 Component Selection

The component choice will depend on many factors including the available production machinery, the cost required, the size and the design cost and timescale. The initial component selections will be based on cost, size and availability of the major integrated circuits involved. Surface Mounted versions of the integrated circuits (see Chapter 4) may, on the face of it, offer size advantages over their leaded counterparts, but the circuit interconnections may be such that almost all of the connections from the IC must be made to components which are likely to be mounted on the opposite side of the board. In this case a leaded device mounted through the board may be the best answer as the SM device will need via holes for most of its pins. These vias may take the overall size for the surface mounted IC package up to the size of its leaded counterpart. The converse will apply to those integrated circuits (e.g., buffers) whose pins virtually all connect to the pins of another integrated circuit. In this case, if the pinning of the devices is suitably matched, the assembly would benefit considerably from the use of two Surface Mounted devices side by side. A number of components will almost certainly be in leaded form for reasons of specification, cost, or availability, and these will therefore be fixed at an early stage. The remaining resistors, capacitors and semiconductors will probably be specified initially as Surface Mounted devices for reasons of cost and size of the final assembly. Some of these may, however, have to be changed to leaded form at the layout stage if interconnection on the two specified layers proves to be impossible with all SMDs.

The example product has now become a Mixprint assembly using a combination of leaded and surface mounted devices on the top surface and surface mounted devices on the underside.

10.4.2 Mechanical Constraints

The mechanical restrictions on the assembly could also include overall thickness and even a limiting height for both sides. If the ultimate minimum thickness is required, then it may be necessary to convert more of the devices to Surface Mounted form as these are generally thinner than their leaded equivalents. This requirement may mean that a double-sided board is no longer possible and a multilayer approach may be needed. This decision would, however, open up further options as the extra layers of interconnect will probably remove the need for track jumping by components and a larger proportion can become Surface Mounted types.

The size and shape of the board may be fully defined mechanically at the outset because it has to fit into a specified area of the total system. In this case the PCB can be specified in detail and the designer will have to select his components to match. Alternatively the system design may be able to cope with a range of board sizes and shapes and, in this situation, the designer may be able to select the optimum aspect ratio for the circuit and components being used. In general, long thin boards, although sometimes necessary, are not the easiest shape to interconnect. In this specific example, because of its cost conscious nature, it has been assumed that sufficient design compromise can be achieved to ensure that a two-layer board layout would be possible without recourse to multilayers.

10.4.3 Board Layout Stage

The layout of the example described above can now start with the required input data of the circuit diagram, the component parts list, and the board size, shape and number of layers. The design could be achieved in many ways ranging from direct taping (black adhesive tape applied to a transparent base film at some magnification), through digitising to full CAD. It will be assumed that a CAD system is available for the layout design of the example. The layout engineer/designer will input the component data in the form of the 'library shapes' of all the devices taken from his machine's memory or, if new devices are to be used, he will first create new shapes specially for this task. A typical 'library shape' is shown in Figure 10.3, together with its computer print-out definition. These 'library shapes' will indicate the maximum body outline sizes, the conductor pad shapes and the numbering sequence of the connection pins. The circuit diagram interconnections may be entered by means of a drawn 'Schematics' circuit, or by a connection listing of all the component interconnections (see Section 5.5.2.3). A partial connection listing is shown in Figure 10.4, illustrating the nature of the input data form used in CAD systems. The board size and shape will normally be applied to the CAD system by effectively drawing it on the screen or by defining the corner co-ordinates. The number of interconnect layers to be used will usually be specified by an input from the keyboard of the CAD.

```
L 16 71 54 28
   3  6 30
   8  6 30
  13  6 30
  18  6 30
  23  6 30
  28  6 30
  33  6 30
  38  6 30
  43  6 30
  48  6 30
  53  6 30
  58  6 30
  63  6 30
  68  6 30
  68 48 30
  63 48 30
  58 48 30
  53 48 30
  48 48 30
  43 48 30
  38 48 30
  33 48 30
  28 48 30
  23 48 30
  18 48 30
  13 48 30
   8 48 30
   3 48 30
```

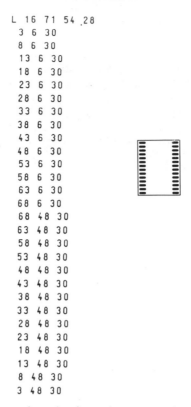

Fig. 10.3 Typical library shape drawing and computer printout for the CAD system.

FROM		TO											
COMP	PIN	COMP	PIN	COMP	PIN	COMP	PIN	COMP	PIN	COMP	PIN	COMP	PIN
IC1	14	D4	1	IC4	8	IC2	14	C12	1	C17	1	R15	1
IC1	4	IC1	5	IC1	12	R19	2	C13	2	C15	2	R18	2
R18	2	C14	2	R16	2	C16	2	IC1	11	R17	2	C12	2
IC1	10	R17	1	J1	4								
IC1	2	R20	2	C16	1	R16	1						
IC1	3	R19	1	R12	2	TR8	1						
IC1	15	R12	1	TR8	2	ZD5	1						
R13	2	TR8	3	R14	2	TR7	1						
IC1	6	C14	1										
IC1	8	R28	1										
ZD5	2	R13	1										

Fig. 10.4 Part of a typical component interconnection listing for the CAD system.

10.4.3.1 LAYOUT DESIGN

The printed circuit board layout designer is now ready to start designing the board. The CAD system being used may be capable of automatic placement of components and automatic routing of interconnections. If this is so then the designer may make use of these facilities for his first attempt at the layout. The machine will then position the components on both sides of the board and interconnect them using its own built-in software programmes to produce a completed board design. In most SMD or Mixprint boards, the present generation of CAD systems are unlikely to be able to achieve a satisfactory solution by this method, and some intervention by the layout designer will almost certainly be required (see Section 5.5.2.3). In the case of this example, the requirement to achieve a high packing density will imply a considerable involvement by the designer in both placement and routing.

The board layout design is an iterative process and the designer will try out a number of component layouts before deciding on one which looks likely to be a probable solution. Having achieved a promising looking component layout using the selected components, he will check the interconnections of the major devices for obvious problem areas (e.g., power tracking) and decide whether to proceed to the tracking stage or try again. After a few modifications to his basic layout, the designer will move on to the tracking stage. He may allow the machine to use its automatic routing procedure to track the board and then tidy up the result, or he may use manual tracking of the entire board, depending upon his reading of the situation. In general, the routing will be done by a combination of man and machine, with a number of options being tried and discarded before the final tracked board is available on the CAD system screen.

The design will now be carefully checked both automatically by the machine and, where possible, manually by the project engineer. A pen-plot (paper drawing of the layout) will probably be produced by the CAD system to facilitate this checking procedure. The checked design is then transferred from the CAD system to a photoplotter for the production of the photographic artworks for the conductor and solder resist patterns and also, in the relevant form, the board drilling, assembly and test information for the various production machines. Any modifications to the original parts list will also be passed to the project engineer so that these data can be updated to take account of any changes from SM to leaded components or *vice versa* made during the board layout phase. This phase of the design task may take a total of 75 man hours of computer time to complete.

The design is now completed and all data necessary to proceed to the next stage are available from the CAD system.

10.5 DESIGN EVALUATION

The system design will have been evaluated at the breadboard stage and the board layout will have been checked both by the CAD system and by the project engineer, but the board must be actually manufactured and assembled as a working device in order to evaluate the final design fully against the requirements of the original specification.

10.5.1   Stages of Sample Manufacture

A number of stages may go to make up this design evaluation depending upon the type of board, its environmental requirements and the particular market for which it is intended. A general situation could include the following stages:

(i)   Make one or two boards for electrical testing.
(ii) Make 6 to 10 boards for customer trials as a complete system to check for any peripheral effects.
(iii) Make a larger batch (50 to 500) of boards for approval testing, probably by an independent or government sponsored test house, or endurance testing to prove suitability for the specified environmental conditions. The manufacturability of the design using the available or specifically obtained production equipment can also be checked by making this batch.
(iv) Make a production batch of limited numbers (e.g., 1000) in order to try out the new board on the system assembly line prior to commitment to full production.

The last two stages—(iii) and (iv)—would require commitment to production tooling as the assemblies would need to be representative of actual production to give meaningful results.

10.5.1.1   STAGE 1—THE FIRST PROTOTYPE

The first stage of prototype manufacture is normally limited to one or two devices and is basically required to check out the board's electrical performance against the specification in respect of normal operation and limit conditions. The boards are unlikely to be subjected to the full environmental conditions and can therefore be made and assembled by 'knife and fork' methods. As only electrical assessment is to be carried out, the assemblies do not have to follow production methods, but only need to be made by the easiest and quickest route. This can be entirely manual, but it is more likely that manual assembly combined with the use of machine soldering (both wave and reflow) would be used. If production is assured and the relevant machinery is readily available, then it may, even at this stage, be worth writing the software programmes needed to carry out the assembly by machine. The example computer board is a double-sided plated-through-hole fibre glass PCB and the first prototypes can be made by a manual method. Solder resist would probably be needed, even for the first boards, in order to define the reflow soldering pads fully for the top surface SMDs. The Mixprint assembly could be made by hand at the one, or two level, but it is probably considerably more convenient to use placement machinery if this is available. Unless the board is completely wrong, which is unlikely, or the project is only an exercise with no production intended, then the cost of programming the machinery will be incurred anyway, and the added convenience for all the following batches makes this method attractive.

The board would be tested electrically in the laboratory to the extremes of the specified power supply voltage with input signal levels and loads set to both maximum and minimum limits, to check the performance of the board against its basic specification. Assuming that no problems were found, then the board would progress to the next stage of manufacture of a larger number of samples for the customers' trials of the completed system. If the board is found to be faulty because of unforeseen electrical layout problems, misunderstandings in the specification, or simple human errors, then the design can be changed on the CAD system and a new sample made, one of the advantages of CAD being the speed at which changes can be made. In the event of board malfunction due to a circuit design problem, then rather deeper investigation of the fault would be needed before modification of the layout.

10.5.1.2 STAGE 2—CUSTOMER EVALUATION DEVICES

The second stage samples would be made and tested as part of the total system to check out the performance under real operating conditions. In these tests the different performance of chip components, with their reduced lead inductance and generally shorter connection paths, will be observed and any changes of value required passed back to the supplier for incorporation in the next stage of samples.

10.5.1.3 STAGE 3—LIFE TEST SAMPLES

The major and most intensive testing and evaluation of the design is carried out on these third-stage samples, made for life, environmental, and step stress testing. This testing can take many forms, and different markets have widely differing requirements with regard to the environment and the number of devices and hours of testing needed to prove the suitability of new assemblies.

Life testing is usually in the form of thermal cycling over the maximum range of the specification with the power and loads at their maximum stress levels. Duration of the test can be from a few hours to thousands of hours with the 'norm' being of the order of 500 hours. A life test cycle for the automotive market could be, for example, 600 hours with full power applied with a thermal cycle of $-35°$ to $+85°C$, completed every 2 hours. A total of more than 300 devices would have to pass this test to prove the suitability of the product. The example board would probably have a less stringent specification as it is a computer system board, not intended for use at extremes of temperature.

Incorrect selection of substrate and components can cause problems for SMAs under lifetest conditions as the small, often leadless components have very little compliance in themselves, and rely in some cases on a reasonable match to overcome thermal stresses without failure (Chapter 2.2.1 and Chapter 9.10).

Environmental testing will include a number of tests generally required to be carried out on a small number of boards. These tests will be designed to assess the suitability of the assembly for operation in the real environment which it will experience in the field.

The test could include the following:

    (i) High temperature storage.
    (ii) Low temperature storage.
   (iii) Humidity.
   (iv) Vibration.
    (v) Bump.
   (vi) Electrical spikes and pulses.
  (vii) Electrical over-voltage fault.
 (viii) Electrical field susceptibility.
   (ix) Electrical field interference.

In some of these tests SMDs will give superior performance to their leaded counterparts, simply by virtue of their smaller size. In others, this small size is a disadvantage and special measures may have to be taken to effectively provide a barrier between the Surface Mounted device and the environmental condition likely to cause the problem. An example is the humidity effect on adjacent high impedance pins of integrated circuits where the DIL device has an advantage merely because of its increased lead spacing.

Step Stress Testing (Chapter 9.10) is a means of obtaining information about the assembly design similar to that obtained with other forms of life testing except that, with Step Stress, the results, in the form of failed assemblies, occur more quickly. This form of testing can be very useful in assessing how much 'in hand' the design has relative to its specified requirement, but care must be taken with the increase in stress so as to avoid unrepresentative failure modes. Warming a newly laid hen's egg can shorten the chick's incubation period. Overheating only results in a cooked egg for breakfast. Publication of Step Stress Test results can also cause problems as 'customers' do not normally expect devices to fail under test. These tests can, however, be invaluable for improvement of the device design by feeding back information of failure modes affected by the assembly process which can then be eliminated.

The making of larger batches also gives the possibility of testing out the design for manufacturability and the suitability of the jigs and tools to meet the required assembly stage times and costs. Minor modifications to the device or the jigs to overcome any problems at this stage can save large sums of money during the large scale production phase.

10.6 MANUFACTURE AND TEST STAGES

The assembly, having passed all the test requirements of its specification, will need to progress to the manufacturing phase. If the example in question is the first Surface Mounted Assembly to be manufactured then some new equipment will need to be purchased and commissioned. More generally, a new design will follow on from other SMAs already in production and a suitable range of manufacturing equipment will already be available. As discussed in Chapter 5 the product design must match the processes that are intended to be used in production. It will be assumed that this has been done and Figure 10.5 illustrates the processes to be used in the manufacture of the example.

10.6.1 Screen Print Solder Cream

The top surface of this example board has a number of surface mounted devices including ICs and a smaller quantity of leaded components. A soldering process is therefore required for these SMDs which will not block the holes needed for later insertion of the leaded devices. Reflow soldering using solder cream would therefore be chosen. As there are a large number of devices involved, screen printing of the solder cream will be the optimum method, rather than the other alternatives described in Chapter 7. A typical solder cream screen printer is shown in Figure 10.6.

In this process all of the solder cream is screen printed onto the board at the same time, using either a mesh screen or a stencil (Section 7.3.2.3). The screen is placed into the printer and solder cream is placed onto its top surface. A board is

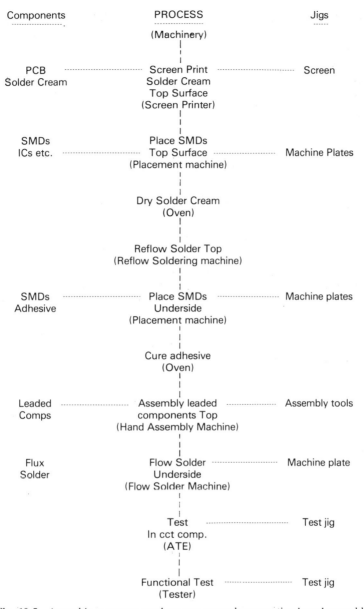

Fig. 10.5 Assembly processes used on our example computing board assembly.

aligned on the machine transfer plate and the plate is pushed into the machine to rest under the screen. The machine 'squeegee' then wipes across the screen, printing the solder paste through the screen mesh onto the board. The machine transfer plate is then withdrawn and the board, with its pattern of screened on solder paste, is removed. (The process using a stencil is identical.)

An investment in the machine is required in order to be able to use the process of screen printing and a specific screen is required for each board to be printed. A facility for cleaning the screens at the end of each run is also required as the solder cream cannot be left on the screen.

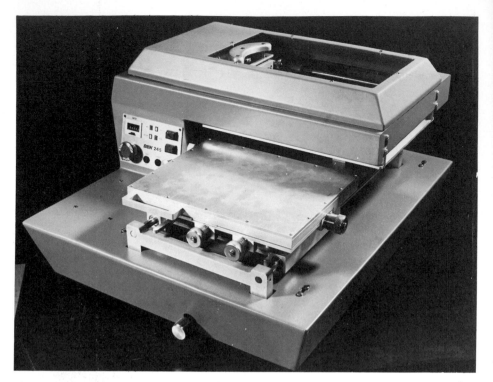

Fig. 10.6 Screen printer used for screening solder cream onto the top surface of the board prior to component placement. (Courtesy of DEK Printing Machines Ltd)

10.6.2 Top Surface SMD Placement

The SMDs on the top surface of the board will be primarily integrated circuits, together with a few of the passive devices and semiconductors. Placement machinery which can handle a wide range of component sizes and shapes will be needed. A glue head is not required as the devices will be placed in solder cream and reflow soldered. It is unlikely that all the components will be available in tape and reel packaging and a machine which can handle a range of tape sizes as well as sticks on vibratory feeders is therefore necessary (see Chapter 6.3.3.2).

The machine would be loaded with the sticks and reels of components and would place the devices onto their correct positions on the board under the control of the software programme entered into the machine. Direct transfer of the board design data from the CAD system can be used in order to generate this software programme, thus eliminating introduced human errors. A specific tooling plate would be required to hold the PCB in the correct position on the machine, and the board would be loaded onto this plate on the machine. The board would be unloaded at the end of the machine placement and a new board loaded for the machine to repeat the placement programme.

10.6.3 Reflow Soldering

The board, together with its solder cream and components, must be passed through an oven prior to reflow soldering in order to dry the solder cream

(Section 7.3.3). If this oven is a belt feed type, then no extra investment in jigs is required and the boards may be passed directly onto the oven belt from the placement machine, thus eliminating the need for expensive racking to store the 'wet' boards.

After the solder cream has been dried, it must be reflowed in order to form the joint between the board and the component. This reflow soldering can be performed by a number of different processes of which the cleanest and most easily controllable is Vapour Phase Reflow (Section 7.3.4.3). An in-line belt feed machine is desirable and then no specific jigs are needed. One example of this type of machine is shown in Figure 10.7. An area of belt or a flat table is required

Fig. 10.7 Vapour Phase Reflow machine used to reflow the screened-on solder cream to form the component/board solder joints. (Courtesy of Hedinair Ltd, England)

for initial storage of the boards after reflow as they generally emerge from the machine at too high a temperature to handle. No masking of the board is required in this process as there is no further solder applied and the vapour used is not normally harmful to the board or its metallic finish. A typical time/temperature profile for an in-line Vapour Phase Reflow System is shown in Figure 10.8.

The board, once cooled, is now in a handleable state with all the top surface SMDs placed and soldered in position. A component level test of the board at this stage would be a practical possibility, but would only be used with very complex and expensive assemblies.

10.6.4 Underside SMD Placement

The Surface Mounted devices to be placed on the underside of the board will be flow soldered at the same time as the leaded devices. They will therefore have to be fixed with adhesive in order to prevent them falling off when the board is inverted prior to leaded component assembly. The components used on the underside of the board are likely to be resistors, capacitors, and semiconductors

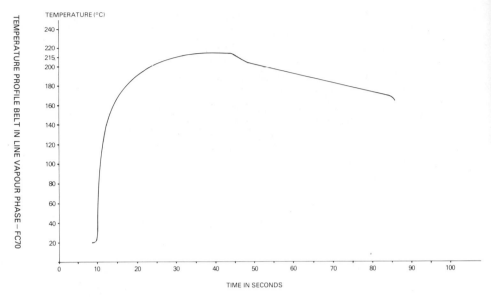

Fig. 10.8 Time versus Temperature graph for Vapour Phase Soldering when using an in-line system.

in the 0805, 1206, SOT23 and Minimelf package styles, and hence will be supplied in 8 mm embossed plastic or metal tape. A fast multihead machine would be suitable for this task (Section 6.3.3.4), the number of heads needed being defined by the actual number of SMDs used. The Philips MCM II (Figure 6.8) machine with three heads, for example, picks up 96 components simultaneously, applies glue to their undersides and then places them in sequence as defined by its software control programme. The example board could have all of its underside SMDs placed in 10 seconds on this type of machine. A set of eight jigs is needed to hold the board at the eight positions of the rotary machine table.

The boards are unloaded from this machine and placed on the conveyor belt of an oven in order to cure the adhesive so that the boards, with their chips attached, can be handled during the subsequent process stages.

10.6.5 Leaded Assembly

The leaded components inserted from the top of the board should be few in number, and can therefore be inserted by hand at assembly stations which may have additional aids to indicate the correct sequence of component insertions to the operator. A typical assembly station of this type is shown in Figure 10.9. Automatic assembly is unlikely to be economic for the example board on its own because, by design, the leaded devices are likely to be few in number and very variable in size and shape. Assembly of the 20 or so leaded devices could take approximately 100 seconds using this equipment.

The investment in these hand assembly stations can be relatively small, especially if the components are formed and cropped prior to insertion on the board.

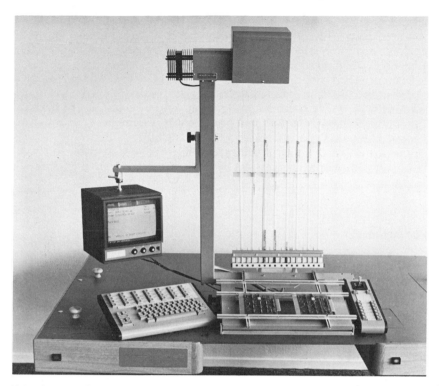

Fig. 10.9 An example of a light guided operator assembly aid for hand leaded component insertion. (Courtesy of Logpoint, Sincotron AB, Sweden)

10.6.6. Flow Soldering

The entire underside of the board containing both Surface Mounted and leaded devices will be soldered in a single pass over a flow solder line using a dual-wave system in order to achieve correct soldering (Section 7.4.4.5). The boards will be held by jigs which pass over the system by means of a conveyor belt. Fluxing, pre-heating and cleaning (underside) are normally included in the total soldering system and no other processes are required externally to achieve full board soldering. The jigs used by the conveyor will normally require inserts to carry specific shaped or sized boards, with sufficient jigs and inserts being required to cope with the throughput time of the system which could be 20 minutes.

10.6.7 Testing

It has been argued that in-circuit component testing by means of Automatic Test Equipment is mainly required to detect the variety of faults found in hand assembled boards, and that, with Surface Mounting being a machine orientated assembly method, it will no longer be required. This may eventually be true when there are few hand assembled components left, the SMDs are so reliable as delivered and the soldering processes are so well controlled that they require no further testing until the assembly undergoes its final functional test. In the meantime, in-circuit component testing using ATE (Section 8.3.1) will give a considerable safeguard against incorrectly assembled modules being powered-up at the functional test stage and inflicting significant damage on the board or even to the tester. In-circuit component testing can also pinpoint the faulty device much more accurately than any other form of testing and is therefore an invaluable repair aid when faults do occur.

The example board with SMDs on both surfaces may need a double-sided jig as some of the top SM ICs may not have connections accessible from the underside of the board. This would mean a larger investment in the test jig for the module, but this cost is dwarfed by the original price of the equipment itself, which can be in excess of $200,000 for a fast and powerful machine.

Every board will also require a dedicated functional test in order to prove that the correct set of components, verified by the ATE, actually do the job they were designed for.

10.6.8 Summary

The efficient manufacture of the example board requires a whole range of machinery and equipment for the combination of processes necessary to complete its assembly. These in turn need tooling and specific jigs or plates to carry the particular printed circuit board required to be made. A few component forming tools will also be required for the remaining leaded devices, although these could be dispensed with if the suppliers could be persuaded to supply the devices in the required form.

An advantage of the manufacturing system outlined above for this particular example is that it is reasonably flexible and could then be used to make a totally different assembly at very little extra cost.

10.6.9 Monitoring of Processes and Reaction to Change

The manufacture of assemblies using the above equipment must be controlled in order to ensure adequate quality at all times. A system of quality audit and inspection is necessary to ensure that the products are of a consistent and acceptable level of quality.

Monitoring of the individual basic processes is the most likely approach which will give consistent quality over a wide range of products. A quality monitoring procedure for each technique should therefore be used to ensure that there is no drifting of the actual process parameters with time. Checks of the process using standard test pieces on a regular basis can give the required information and the results obtained can be accurately compared with previous ones. Examples of the photographic masters of three test boards used for this task are shown in Figure 10.10.

Many products will require design alterations during their life cycle, and a flexible production assembly area will allow changes to be implemented with the minimum of fuss and disruption. Flexible software controlled machines should therefore be used to allow quick and, if necessary, frequent changes of programme. At the same time a disciplined and controlled approach will ensure that there is no mix-up of the assemblies. A production area set-up to handle a minimum batch size of one with computer control of all the machines, using, for example, a bar code on the actual assembly is the ideal target at which to aim.

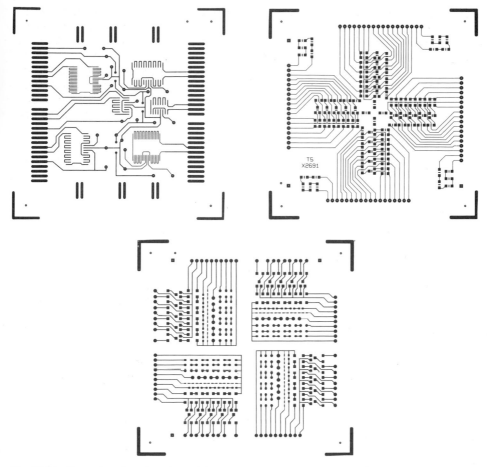

Fig. 10.10 Examples of test boards used for process parameter definition and checking. (Courtesy of Mullard)

10.7 FIELD PERFORMANCE

This is the final stage in a product's life cycle, but the most important. All previous stages have had the one objective of getting the product sold and into the field to be used by the customer.

10.7.1 Reliability

The ultimate achievement is that once sold the product always performs the desired function, never requires maintaining, never goes wrong or needs repairing. For many products involving moving mechanical parts, high voltage or current switching this is too ambitious. At the very least, regular maintenance in the form of cleaning and lubrication is necessary. However, for electronic sub-assemblies utilising SMA, this ambition should often be achievable provided the field environment is correctly anticipated, a toleranced design is used and manufacturing processes are adequately controlled. It will still not be invariably achieved. Inspection and test are never one hundred per cent efficient;

components from suppliers, particularly integrated circuits, can exhibit failure mechanisms and tarnished conductors will invite poor solder joints. Nevertheless the component reliabilities of the passive components (referred to in Sections 3.4.2 and 3.5.3), having failure rates of $<1 \times 10^{-9}$ per component hour, the active components which can have failure rates as low as 10×10^{-9} for transistors (Section 4.3.9) and $7 \cdot 8 \times 10^{-6}$ (Section 4.4.5) for integrated circuits when combined with compatible substrates (Section 2.2.1) can achieve excellent reliability performance. This compatibility, particularly with regard to the thermal coefficient of expansion, is again discussed in Section 9.10.2.

The overall effect on field performance through the use of SMA should be beneficial. Component reliabilities are as good as, or better than, wire leaded alternatives and their mass is lower. Placement and soldering processes are capable of tight control. The potential problems are in thermal cycling if unwise decisions are taken with unsuitable substrate materials and incompatible components, and the dangers of inflicting damage during in-circuit component testing with the highest component packing densities.

10.7.2 Repair

Whilst always aiming to design a product to be so reliable as not to require repairing, it would be foolish to produce one that was irrepairable. Repair will be necessary during the manufacturing process and this aspect was considered in Section 8.4. What of repair in the field? The replacement of a simple surface mounted component is no more difficult than for a wire-ended device. Replacement of an 84-pin chip carrier or flat pack will require more skill. The main problem, however, with any complex electronic system is to determine the source of malfunction and to prove, in due course, after effecting the physical repair, that the fault has been rectified. To do this, reliably, generally requires bulky and expensive test equipment. Increasingly, therefore, satisfactory fault analysis and rectification will demand either that the equipment is returned to the original manufacturer, or alternatively that only board replacement takes place in the field followed by board repair at the original supplier's premises. This philosophy is not unique to the application of SMA, but its ability to achieve greater packing densities will tend to encourage more complex circuits per substrate and more complex systems overall and hence reinforce this philosophy.

10.8 PRODUCT UPGRADING

The timescale between foreseeing a market requirement, defining a product and launching it into the market, may be several years. Once in the field, its performance may be overtaken by a competitor's. Sometimes, the original product can be improved by adding new software without recourse to any hardware changes. Often, however, this is not possible and perhaps mechanical alterations demand changes to the electronic hardware. Such changes should be easy to handle with modern computer-aided design facilities and computer controlled manufacturing machines. CAD designed substrate layouts can be altered in hours or at most days and new software to control SMD assembly machines can be edited as quickly. Provided no new processes are demanded, a re-vamped product design can be introduced very rapidly. The main timescale limitations are likely to be in obtaining new components in production quantities and in custom jigs for testing. The need for these custom jigs is likely if in-circuit component testing is employed because of the close packing densities on the substrate, preventing the use of general purpose test jigs (Section 8.3.1). Otherwise, provided production machinery has been chosen wisely for its flexibility, product design changes can be brought about very rapidly. Perhaps the launch of the future new SMA model will be held up by the late arrival of its new label, its packing, or instruction manual, rather than in modifying the design or changing over the production line.

Chapter 11

THE FUTURE

J. F. PAWLING and G. A. WILLARD

Mullard Mitcham, Mitcham, Surrey, England

11.1 THE FUTURE

The future of SMA is linked with technological advances in silicon integrated circuits and the demand for greater complexity from the market place. As the number of transistor gates per square millimetre of silicon continues to expand at the current rate of doubling every one or two years, there will always be applications that can use the increased complexity offered by the IC. This trend will increase the number of input and output ports around the chip. It has been suggested that by 1990 25% of silicon integrated circuits will have greater than 40 connection pins[1]. These more complex circuits will be required to work at faster speeds within the chip and this, in turn, will require reduced lead spacings and shorter conductor paths. The resulting pressure for greater packing densities will demand finer line widths for connection patterns and smaller peripheral components to reduce the area of substrate used. This compression will increase the interconnection problems which can only be solved by increasing the number of interconnect layers. These layers need to be linked by vertical 'via' connections which, in order not to use up significant substrate area, will have to become smaller in cross-sectional area and 'blind' or 'buried' so as not to take up space on every layer. Nevertheless, however much contraction of size is technologically possible in the two dimensions of a substrate, it will also become increasingly necessary to extend the electronics assembly into three dimensions. Greater packing densities and faster operating speeds will also increase the dissipation problems from internally generated heat. Another way of regarding this is that SMA has so far only offered solutions for the low level logic and signal processing circuits. A comparable technological advance is necessary for the larger power dissipation stages both in component encapsulation, substrate materials and assembly machinery. In the following sections the above points will be considered in greater detail and some possible solutions discussed.

11.2 GREATER PACKING DENSITIES

11.2.1 Two Dimensions

11.2.1.1 THE SUBSTRATE

The trend will be for integrated circuits having higher pin counts to be packed as close together as possible. To interconnect such circuits successfully will require finer conductor line widths and inter-layer 'vias' which use the minimum of substrate area.

If one first considers the conductor line width, then these can be generated by an 'additive process' or a 'subtractive' one, the most common subtractive process being the etched copper foil laminated printed circuit board. To reduce the line widths with this technique, one must also progressively reduce the thickness of the copper foil. The most common PCB material uses 1 oz copper, i.e., a foil thickness of 35 microns ($1 \cdot 4$ thousandths of an inch) and sets a limit of about 10 thousandths (250 microns) in line width. A $17 \cdot 5$ micron (½ oz) foil enables line widths of 150 microns to be achieved. As the foil becomes thinner it becomes difficult to handle during the laminate manufacturing process, particularly below about 12 microns. Printed circuit boards can however be produced by 'additive' methods. Thin conductors are then less of a problem and conductor line widths of 50 microns are possible. An additional advantage with this process is that the conductors have vertical straight sides with no under cut which, with etched boards, can be the cause of poor efficiency.

Additive processes have been used for many years on alumina substrates. It is possible to use the screen printing techniques employed in thick film hybrids (Chapter 2.3.2.9), with polymer-based conductor inks on fibre glass epoxy substrate materials. Copper based inks are attractive, being inexpensive, and, unlike noble metals such as gold and silver, do not diffuse into the molten solder. The prime problem of tarnishing is less than with stoved thick film inks and may be further reduced by the use of vapour phase reflow soldering, which eliminates the presence of oxygen when the substrates are raised to the soldering temperatures. Multilayers can be produced by alternate printing of conductors and insulating layers, and substrate materials can cover a range of plastics and metals as well as the more conventional hybrid ones, such as alumina and beryllia.

When designing the substrate layout, once a few connections are completed radiating from a component such as an IC, others are only possible by jumping across those already made. To do this requires either a convenient component body or sooner or later a descent to a different interconnection level. This transport to a different level requires a 'via' connection. In a multilayer printed circuit board these 'vias' are produced by plating the walls of a hole drilled in the substrate. There are mechanical constraints, limiting how small in diameter such a hole can be, and hence how much area the resulting 'via' consumes of the substrate. These include the thickness of the substrate insulation material (the ratio of substrate thickness to hole diameter is usually 3:1), and how fine a drill can be used without breaking. Unless care is taken, attempts to achieve greater packing densities are frustrated by the size and number of 'vias' required to achieve electrical interconnection.

The additive processes described earlier can assist with this 'via' problem. The ratio of substrate thickness to hole diameter can be 6:1 or more, and if polymer inks are used, the 'via' electrical connection can be achieved by squeezing the ink through the holes in the substrate. The holes can also be punched provided the substrate material is 'punchable' and the quantities are large enough. The substrate need not be confined to epoxy resin but can be metal cored, plastic or resin paper. The smallest holes can probably be produced by a high power carbon dioxide laser, instead of a conventional drilling machine, as the popular printed circuit board materials, such as glass and epoxy resins, are good absorbers of radiation at the operating wavelengths of these devices. Holes are said to be producible as small as 5 thousandths of an inch^2. With laminated materials there can be a problem using lasers due to the reflection from the copper, but this is eliminated when the conductor material is added later. In summary, to gain the greatest packing densities in two dimensions, the trends will be towards thinner conductor layers enabling finer conductor line widths and minimum area 'vias' to be achieved. The thin conductor layers are increasingly likely to be introduced by an additive process and the small 'via' holes produced by laser rather than conventional drilling.

Multilayer interconnects for SMA can also be achieved by the use of 'Multiwire'[3]. With this technique, thin insulated copper wires providing component interconnection are laid onto a special adhesive material which, itself, is bonded to a normal epoxy glass laminate with etched copper surfaces. Advantages of the technique are high component packing densities, short lead times from specification to manufacture, controlled signal line impedance, higher reliability and lower cost. The whole process is computer-aided, both in the design and manufacturing phases. The software package determines the optimum wire routing on the basis of minimum length, to satisfy the requirements of the circuit, the component layout and the board outline.

The thin insulated wires are able to cross one another and are themselves encapsulated in thin layers on the surface of the carrier substrate. All holes are plated through, providing both the electrical and mechanical connection between the wire and the hole, and to the power or ground planes, which are normally provided by etched copper foils bonded to the surface of the base material. Pads formed on the surface of the wiring layers facilitate soldering of the components (Figure 11.1).

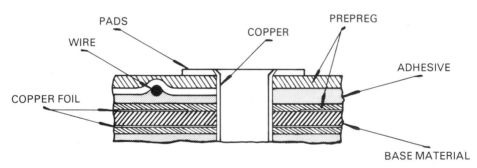

Fig. 11.1   Cross-section of Multiwire board. (Courtesy of Kollmorgen Corporation)

Microwire is an adaptation of the basic Multiwire system using thinner wires to produce even more densely packed assemblies.

A number of other approaches have been proposed, aimed at achieving a suitable substrate plane for mounting and interconnecting surface mounted devices in a complex interconnection pattern. These newer concepts of interconnection will find areas of application where the highest density is required and some are likely to find a wider acceptance if the costs can be reduced to a level where large scale usage can be contemplated. Many of the possible methods will, however, be too costly to find widespread acceptance and will fall by the wayside or be limited to the small quantity, expensive areas of the total market. The overall trend must be towards denser packing of these very small surface mounted devices in order to take full advantage of the size reductions which become possible. This will inevitably lead to some of the currently more expensive multilayer substrate systems eventually becoming the 'norm' and effectively replacing the double-sided PTH board as the standard substrate.

11.2.1.2 THE COMPONENTS

A major proportion of the substrate area of a complex electronics system is usually taken up by the silicon integrated circuits. This is particularly true when using the DIL package but still relevant when considering SO and Chip Carrier encapsulations. Much of the surface area of an IC package is employed in bringing out connections from the silicon chip and not by the chip itself. Greater packing densities can be achieved by reducing the number of integrated circuits used, i.e., by combining several into one chip. This customised IC design approach, as it becomes more popular, is also likely to be extended to defining the required mechanical encapsulation. Care in specifying the circuit pin order can ease the interconnection problems between one IC and another and hence save substrate area, as can the use of three IC 'encapsulations' which are particularly miserly in their use of substrate area, i.e., Flip chip, Tape Automated Bonding (TAB)[4], and Epoxy Plastic Integrated Circuit (EPIC)[5].

The Flip chip is supplied by the manufacturer with connection bumps, generally of solder, directly on the surface of the semiconductor die, which is then flipped over and reflow soldered onto the substrate face downwards. It requires a rigid substrate such as ceramic and has been used by Hewlett Packard and promoted by IBM. The TAB approach is ideal for large volume production and, by offering a flexible connection from the silicon, can be used on more flexible substrate materials such as printed circuit boards. The ICs are supplied in a manner reminiscent of chip resistors and capacitors, i.e., on a tape reel. The tape consists of bonded copper foil and polyimide, the connection 'fingers' to the silicon chip being etched in the copper. These are thermo-compression bonded to the chip after gold plating and then cut free at the far ends once the chip is positioned correctly on the final 'mother' printed circuit board. The 'EPIC' is a low cost carrier approach proposed by British Telecom, using glass epoxy laminate as the carrier material. Silicon chips are bonded to one layer of the laminate and connections brought out via plated-through holes to the other surface. The devices are batch produced together, and then separated by slicing through the plated-through holes. The half tubes so formed then act as the contacts to the outside world.

11.2.2 Three Dimensions

However much contraction of size is technologically possible in one plane, ultimate miniaturisation will be achieved by using all three dimensions. Electronics in the pre-printed circuit board era used all three and it was the introduction of the PCB and hybrids that took electronics assemblies down the two dimensional route.

Some of the technological advance employed to increase the packing density in two dimensions will, however, also assist when considering three dimensional executions.

Smaller components and thinner conductor layers, when combined with thin and flexible substrate materials, suggest that, although still desirable to manufacture in planar form, once assembled, soldered and tested, the assembly can be re-shaped in a variety of ways.

Where the final product is encapsulated or boxed, a flexible unit can be concertinaed as shown in Figure 11.2. Alternatively it can be spiralled into a

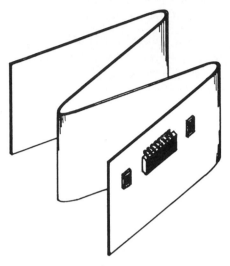

Fig. 11.2 An example of reshaping a flexible two-dimensional substrate into a three-dimensional concertina.

'swiss roll' shape as illustrated in Figure 11.3. The former approach can be combined with employing the outer case and jig as a heat-sink for power devices electrically connected by the edges of the foil PCB. A further possibility is for each folded section to be alternately flexible and rigid. The rigid portion can be laminated to the flexible part so that the connecting paths are continuous and the rigid material can either be a heat sinking metal or a rigid printed circuit board. By electrically connecting, for example, a two layer flexible and two layer rigid PCB, a multilayer is obtained. A limitation to this approach is that circuits, when 'unfolded', are long and narrow and many circuit diagrams do not lend themselves to a strip layout. This can be partly overcome by connecting adjacent

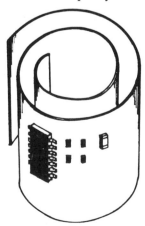

Fig. 11.3 Reshaping a flexible substrate assembly into a 'swiss roll'.

layers when folded either by individual connector pins or by another rigid circuit board mating on the edge of the concertina. A limitation is that not all four surfaces, i.e., both sides of the rigid and flexible prints, can be used for mounting components if two of the surfaces are mated together! However, parts of the rigid board can be removed so that components can be mounted on both sides of the flexible in those areas. This technique has been used by Japanese manufacturers in the design of portable equipment such as video cameras.

A more elementary three-dimensional layout can be achieved by the application of Single-in-Line (SIL) and Dual-in-Line (DIL) daughter substrates. With complex systems requiring high packing densities, the use of simple circuitry daughter modules, mounted either at right angles to the mother board as a SIL or parallel as a DIL, can be most useful. They can be used to mount passive components associated with integrated circuits mounted on the main board—for example filter capacitors or series input resistors. These are illustrated in Figure 11.4.

Fig. 11.4   Single-in-Line (SIL) and Dual-in-Line (DIL) daughter substrates. (Courtesy of Mullard Ltd)

An alternative method of three-dimensional interconnection is used in the Chiprack[6] assembly system marketed by Dowty Electronic Interconnect. In this device, VLSI products encapsulated in the leadless chip style of semiconductor package are interconnected as a three-dimensional assembly. A new style of carrier accommodates the semiconductor die which is connected to pads on either the upper or lower surface of the carrier. Edge connecting racking panels, which are used on all four sides of the assembly, interconnect the carriers and space them one above the other (Figure 11.5). The three dimensional assembly method gives short interconnection paths between the chips, thus reducing propagation delays. Cooling of the system is achieved by means of the skeletal structure of the rack which permits cooling from both sides of the stack. Testing is achieved by probing the contacts on the outside of the rack. Substrate area is considerably reduced (e.g., Z80 prototype system (Figure 11.6) occupies only 20% of the

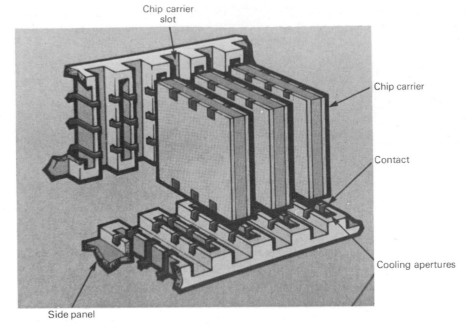

Chip carrier
slot

Chip carrier

Contact

Cooling apertures

Side panel

Fig. 11.5   Interconnection racking of Dowty 'Chiprack' assembly. (Courtesy of Dowty Electronic Interconnect)

Fig. 11.6   Prototype Z80 chiprack assembly. (Courtesy of Dowty Electronic Interconnect)

volume of a DIL system) and the PCB interconnection problems are virtually eliminated. The Chiprack structure would generally be used together with the other components of the complete system on a conventional or Mixprint PCB.

11.2.3 Heat Dissipation

Thermal energy is dissipated, particularly in complex fast switching integrated circuits, in some analogue ICs, and in discrete semiconductors which switch power, or drive displays and motors. SMA has largely concentrated on low level signals where thermal dissipation is low. As packing densities become greater, so do the problems of dissipating the heat generated by those devices mentioned above. The heat must be moved from the massed circuitry by good thermal conductors such as alumina or copper to be finally dissipated into the ambient. Two techniques are likely to gain favour. These are to use a metal based substrate, or to separate the functions of electrical conduction and connection from those of mechanical support and thermal conduction.

11.2.3.1 METAL SUBSTRATES

The prime requirement is a good thermal conducting metal core with an adequate and reliable electrical insulating coating which, at the same time, is so thin as not to prevent heat conducting away from the surface-mounted device. The insulator must also be capable of being coated with the electrical circuit connection pattern.

The following list is not exhaustive but gives some of the possible solutions.

1 Aluminium, steel, or invar-copper substrate with polyimide outer layers[7].
2 Porcelain enamelled steel (PEST) with thick film conductors[8].
3 Glass-ceramic coated metal substrate with thick film conductors[9].
4 Copper, copper/invar or aluminium cored sandwich of epoxy-glass[10].

The power dissipating devices need to be mounted in intimate contact with the surface of the substrate to maintain as low a thermal resistance as possible. The best method is to back bond the naked semiconductor chip to the substrate.

Alternatively, a special surface-mountable fully encapsulated package can be used. For the majority of users the latter approach is to be preferred, particularly if the power device can be assembled rapidly and cheaply on the substrate at a rate comparable to the small signal surface-mountable devices. Machines need to be developed to do this. Because the number of power devices is numerically less than that of the small signal devices, single head machines will generally suffice.

For most circuits, while electrical connection to the small signal drive and logic circuitry will be necessary, thermal linking will be undesirable. Where a flexible PCB is stuck on top of the metal substrate (1 above), continuation of the flexible print will supply the solution. Where the flexible print is not part of the power handling laminate, it can still be used as the connection from the power circuit to the low power areas.

11.2.3.2 SEPARATED ELECTRICAL AND THERMAL CONDUCTION

Previous sections of this chapter have advocated the use of flexible printed circuit substrates with thin copper foils and small laser drilled holes. A proposal

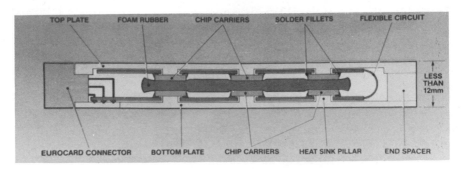

Fig. 11.7   Flexwel Assembly showing the flexible printed circuit, heat sink, pillars and top clamping plates. (Courtesy of Welwyn Microcircuits, Crystalate Electronics Ltd)

to combine the advantages of such an approach with good thermal management has been made by Welwyn and given the name Flexwel[2]. Figure 11.7 shows the main constituent parts of a Flexwel assembly. The chip carrier integrated circuits have been assembled onto a thin polyimide printed circuit. This material is chosen because its elastic modulus is approximately one-sixth that of epoxy-glass. Furthermore, a typical 4-conductor layer circuit will be 15-20 thousandths thick rather than 60 for a rigid board. The combined effect of these two factors is that the flexible circuit has a 'stiffness factor' 1/20th less than a rigid board, thus yielding to any imposed forces in the X-Y plane due to thermal expansion mismatch. The circuit layout is arranged so that there are no conductors beneath the chip carriers. A hole is then possible in the printed circuit which aligns with a cylindrical pillar standing out from the aluminium heatsink. Mechanical clamping of the chip carrier surface to the pillar is later carried out in assembly by the metal top plates. A thermally conducting grease is previously applied to the top of the pillar to ensure a good thermal contact with the chip carrier and also to provide a lubricant for any lateral sliding action promoted by any thermal strain. The assembly can be folded back using a foam rubber centre spacer to provide a two-tier arrangement as illustrated. The proposal offers all the merits attributed to SMA and flexible circuits plus a final rigid and easily handleable assembly with good thermal heat sinking. It is particularly applicable to circuits having an even distribution of power dissipating devices over the whole area of the substrate.

The trend towards three dimensional packaging coupled with the use of surface mounted devices packed closely together increases the problems of cooling of the assembly. Folded flexibles or flex-rigid boards can pack a large amount of computing power into a small three dimensional space but as a result generating heat which must be removed if good quality and reliability are to be achieved. Heat dissipating substrates may give an answer to this problem in the short term, but the point is likely to be reached where normal conventional cooling methods are no longer adequate.

Heat management, at the design stage, will become essential and the conventional methods of conduction, convection or radiation to the surrounding air may need to be replaced with forced cooling. The use of 'finned' substrates and stacks of flexible boards with cooling slots will need to be considered. The extension of these concepts to cooling by means of fans to move the air or even total immersion in fluids will probably not be considered unusual in a few years time.

11.3 DESIGN CONSIDERATIONS

The availability of computer-aided design tools will inevitably lag behind the development of technology. Suppliers are interested in the majority market and not the vanguard who are evolving tomorrow's processes. As systems become more complex, packing densities become greater, layers of interconnect increase and the need for CAD becomes ever more important. As assemblies move from a planar to a three-dimensional world, so the 3D mechanical design concepts of CAD need to be married to the networking algorithms of PCB layout. Furthermore, the simulation of the thermal performance of an assembly will become increasingly necessary at the layout design stage[10].

11.4 SUMMARY

Market pressures will be towards silicon integrated circuits with larger pin counts, substrate designs with greater packing densities, and more thermal dissipation in a given volume. The tools used to meet these demands will include the use of thinner conductor layers on thinner insulating substrates and greater layers of conductors, interconnected by finer 'vias' produced by laser generated holes. Integrated circuit packages will become more space efficient with the silicon chip being proportionately a much greater part. Heat will be removed as soon as possible from the device generating the heat, either by combining the electrical circuit with a metal core, or by separating the functions of electrical conduction and thermal conduction/mechanical support.

REFERENCES

1 BPA Information Services. Technology Report (1984).
2 Kirby, P., 'A Flexible Approach to Chip Carrier Mounting', *Electronic Engineering*.
3 'Multiwire—An Alternative Approach', *New Electronics*, p. 30, 18th October (1983).
4 Small, D. J. and Blain, A. A., 'Tape Automated Bonding', *Brit. Telecom Technol. J.*, No. 3, p. 86, July (1985).
5 Sinnadurai, F. N. and Small, D. J., 'The Manufacture and Reliability of the EPIC Chip Carrier', *Circuit World*, **Vol. 10**, No. 4, p. 30 (1984).
6 Austey, M., 'A Three-dimensional Interconnection System', *Electronic Engineering*, p. 125, March (1985).
7 Wright, R. W., 'Polymer/Metal Substrates for Surface-mounted Devices', Proceedings 2nd Annual IEPS Conference, p. 445 (1982).
8 Published Data. Ferro Elpor Substrates. Ferro (Great Britain) Ltd.
9 Information Sheet No. 1, Wade 'Keralloy'. Wade (Advanced Ceramics) Ltd, England.
10 Kumar, C., 'Thermal Simulation of Printed Circuit Boards', *Electronic Product Design*, p. 39, June (1985).

BIBLIOGRAPHY

1 Topfer, M. L., 'Thick Film Microelectronics', Van Nostrand Reinhold Co.
2 'Surface Mounted Technology', The International Society for Hybrid Microelectronics, Silver Springs, Maryland, USA.

Author Index

(See also separate Subject Index)

N.B. *The figure in brackets following the page number refers to the item number on that page*

Austey, M. 255(6)

Bear, J. 33(14)
Blain, A. A. 255(4)
Brierley, C. J. 226(14)
Brown, A. F. 33(11)
Brunetti, C. 14(6)

Carr, R. 184(15)
Cobb, J. 198(3)
Comerford, F. 184(22)
Cong, L. 33(6)
Cook, S. 198(8)
Cosens, M. E. 198(13)
Costello, J. 198(6)
Cummings, J. P. 33(5)
Curtis, R. W. 14(6)

Dow, S. J. 183(4,5)
Dummer, G. W. A. 14(bibliog.)

Estes, R. 184(19)

Freakes, A. 184(21)

Hagemann, H. J. 14(4)
Hennings, D. 14(4)
Hey, D. 184(8)
Holmes, P. J. 14(bibliog.), 33(4)
Huang, C. 33(6)

Jones, R. D. 33(3)

Kirby, P. 112(19), 255(2)
Klein Wassink, R. J. 183(2), 198(4)
Kozel, B. Roos- 183(1)
Kulesza, F. 184(19)
Kumar, C. 255(10)

Lambert, L. 184(8)
Lawrence, B. 198(7)
Leabach, G. E. 33(16)
Leonida, G. 14(bibliog.)
Lepagnol, J. H. 33(7)
Loasby, R. G. 14(bibliog.), 33(4)

McMillan, I. D. 183(6)
Malhotra, A. K. 33(16)
Manfield, H. G. 33(12)
Mansveld, J. F. 33(9)
Marston, P. 111(14)
Morrisey, J. 198(3)

Otten, 14(5)
Owen, W. 184(16)

Pepper, P. 14(3)
Phillips, L. S. 14(bibliog.)
Pitkanen, D. E. 33(5)
Planar, G. V. 14(bibliog.)

Rahn, A. 184(16)
Richardson, J. 33(1)
Rizzo, J. A. 198(11)
Roberts, L. 184(11)
Roos-Kozel, B. 183(1)
Ross, W.McL. 33(8)

Sargrove, J. A. 14(2)
Schmickl, 14(5)
Sinnadurai, F. N. 255(5)
Small, D. J. 255(4,5)
Smith, M. 198(8)
Snelling, E. C. 14(bibliog.)
Speerschneider, C. J. 33(5)
Spencer, A. 77(27)
Stakhorst, 14(5)
Stein, S. J. 33(6)
Stokinges, H. E. 33(2)
Straw, J. J. 33(16)

Topfer, M. L. 255(bibliog.)

Wagner, G. R. (33)16
Walton, A. 184(14)
Wargold, I. 198(12)
Wassink, R. J. Klein 183(2), 198(4)
Wernicke, R. 14(4)
White, C. E. T. 33(7)
Wood, E. R. 226(11)
Wright, R. W. 33(15), 255(7)

Subject Index

(See also separate Author Index)

Acceptance Quality Levels 93-4
acrylate adhesives 173
active devices 78-112
  history 8
additive processes
  substrates 28-9, 247
adhesives
  clearances in SMDs 36, 174-5
  component attachment 173-6
    inspection 188
  conductive 90, 140-41, 182
  placement machine functions 140-41
  use in wave soldering 119
adjustable inductors 62
ageing
  ceramic capacitors 53-4
Air Knife
  soldering 179-80
alloys
  solder pastes 165
alumina
  substrates 19-21
aluminium
  bonding
    semiconductor devices 92
  electrolytic capacitors 6-7, 56-7
aluminium nitride
  substrates 22
Amistar Corporation
  placement machines 152
artwork 125
assembly (see also manufacturing processes)
  of components 138-62
attachment of components 163-84
automatic inspection
  optical 186, 188
  solder joints 191
automatic testing 192-4, 241-2
availability
  passive devices 75-6
  semiconductor devices 111
axial components
  PCBs 4

balling
  solder paste 167, 190

'bed of nails' test jig 192-3
belt soldering 239
beryllia
  substrates 19, 21-2
'blister' tape 37-8
bonding
  semiconductor devices 92
bow
  substrates 18
bowl feeders 141-2
bridging
  dip soldering 177
  solder inspection 187, 190
  use of Air Knife 180

CAD 116 125-9, 231-2, 233, 255
capacitors
  electrolytic 56-9
  history 5-7, 12
  multilayer 46-54
  polyester 55
  soldering 'footprints' 130-31
  trimmer capacitors 66-7
capital depreciation
  cost comparisons 213
carbon composition resistors 5
cartridge feeders 142
ceramic capacitors
  history 5-6
  multilayer 46-54
ceramic filters 62-3
ceramic IC packages 99
ceramic substrates 19-27
Cerleds 110
chip carriers 98-100, 104
  elastomeric connectors 73
  for crystal oscillators 64, 65
'chip-foil' capacitors 55-6
chip resistors
  component/substrate compatibility 16
Chiprack assembly system 251-2
Chipstrates 30-31
cleaning
  after soldering 182-3
  and coating performance 196
coatings 195-8

PCB manufacture 31
coefficients
  Tce 215
  temperature
    resistors 43-4
    dielectrics
      capacitors 52
  voltage
    resistors 44
compatibility
  components/substrate 15, 16, 215-16
  machine/substrate 17-18
components
  active 78-112
  and packing density 249
  attachment 163-84
  coding 42
  compatibility with substrate 15, 16, 215-16
  passive 34-77
  placement 138-62, 167, 238, 239-40
    inspection 187-9
  selection 230-31
computer-aided design 116, 125-9, 231-2,
    233, 255
computer-aided inspection 186, 188
computer-aided testing 192
condensation soldering 169-71, 239, 240
conduction
  electrical separated from
      thermal 253-4
  reflow heating methods 167-8
conductive adhesives 90, 140-41, 182
conductors (*see also* metallisation)
  dimensions 29, 247
conformal coatings 196-8
connections
  and packing density 247, 248, 251-2
  design 231-2
connectors 68-75
cooling 253-4
copper clad Invar
  and thermal mismatch 215
  heat dissipation 253
copper foil
  substrates 28, 247
costs
  comparisons 210-13
cracking
  solder
    inspection 190
    reliability testing 218-9, 220-25
creams, solder *see* solder pastes
crystals 63-5
curing
  adhesives 143, 173-4, 175-6
cycling tests 214-15, 216-25
cylindrical diodes 87-9

depreciation
  capital cost comparisons 213
design 113-37, 229-36, 255
  advantages of SMA 203-6, 209-10

cost comparisons 213
desoldering
  repair work 194-5
dielectric layers
  capacitors 48-9. 50-54
  thick film substrates 27
dielectric losses
  capacitors 53
dielectric strength
  capacitors 53
digitisers 125
DIL(DIP) packages
  and joint cracking 220-22
  3-dimensional layouts 251
  trends 13
dimensions
  resistors 39
  MELF capacitors 55
  metallised film capacitors 55
  multilayer capacitors 47-8
  SOD-80 88
  SOIC packages 96, 97
  SOT-23 84
  SOT-89 86
  SOT-143 87
  tantalum capacitors 58
diodes 87-9, 109-11
dip coating 196
dip soldering 176-7
discrete semiconductor devices 82-94
dispensing
  adhesives 174-5
  solder pastes 166
dissipation, thermal 253-4
double wave soldering 178-9
Dowty Chiprack assembly system 251-2
drag soldering 177
drilling
  holes
    and packing density 247-8
    ceramic substrates 23-4
drying
  solder pastes 167
Dyna/Pert
  placement machines 147-8, 152-3

edge location
  substrates 17-18
elastomeric connectors 72-3
electrical testing 192-4
  substrate inspection 186
electrolytic capacitors 56-9
  history 6-7
electroplating *see* plating
embossed tape
  packaging 37-8
encapsulation 198
  history 8, 10-11, 79
environmental testing 235-6
EPIC encapsulation 104, 249
epoxy/glass boards 28
epoxy resins
  component attachment 173

conductive adhesives 182
evaluation
  of design 233-6
Excellon
  placement machines 148-9

failure rate
  semiconductor devices 93-4
fault-finding 194-5
feeding methods
  component placement 141-2
ferrite pot core 12
film capacitors 55-6
  history 7
film resistors 5
filters, ceramic 62-3
'Finstrate' 32
FITS (Failures in Time Standard) 93-4
flatness
  substrates 18
flatpacks 100-102, 108
Flex-rigid Prints 71
flexible printed circuits 31, 250-51
flexure cycling 218, 219, 220-25
Flexwell 106, 254
Flip chips 249
flow soldering 176-81, 241
  adhesive clearances 36
fluidised bed methods
  coatings 197-8
fluxes
  cleaning 183
  residues
    inspection 190
foil capacitors
  history 7
  metallised polyester 55
'footprints'
  design rules 121-5
Fuji Machine Manufacturing Co.
  placement machines 153

gas soldering 169
glue *see* adhesives
gold
  bonding
    semiconductor devices 9
grid arrays 102-4

hand assembly
  components 138-9, 144
hand soldering 181-2
hardware
  programming
    placement machines 142, 144, 159-61
health hazards *see* safety
    precautions
heat *see headings beginning*
    thermal . . .
heating methods
  reflow soldering 167-72

holes (*see also* plated-through
    holes)
  location methods
    substrates 18
hot air knives
  soldering 179-80
hot air levelling 172
hot gas soldering 169
humidity tests 215
hybrid circuits
  comparison with SMA 206-8
  history 9-10

ICs *see* integrated circuits
in-circuit testing 192-4, 241-2
inductors 59-62
  history 11-12
infra-red soldering 168-9
inserted components
  computer-aided design 126-8
  PCBs 4
inspection 185-92
insulation
  thick film substrates 27
insulation resistance
  capacitors 53
integrated circuits 94-109
  and packing density 249
  silicon 8
  thermionic valves 1-2
interconnections
  and packing density 247, 248, 251-2
  design 231-2
Invar
  and thermal mismatch 215

JEDEC Standards 98
jet wave soldering 177-9, 181
joints
  soldering 181
    inspection 189-92
    testing 218-19, 220-25

keyswitches 75

labour costs
  comparisons 212-13
Lambda Wave soldering 178-9
laminates
  properties 28
lands *see* pads
laser drilling 247-8
laser scanning
  inspection methods 191
laser soldering 169
laser trimming
  resistors 26-7
layout design *see* design
leaching
  noble metals
    thick film substrates 26
    solder cream 164
leaded chip carriers 99, 101, 137

leaded components
   assembly 240-41
leadless chip carriers 98-100
leadless devices
   misnomer 36
leadless grid arrays 102-4
LEDs (light emitting diodes) 109-11
life testing 235
locating methods
   substrates 17-18
Lockfit transistor 8

machine assembly 139-61
   design restrictions 116-21
   substrate compatibility 17-18
Mamiya Denshi
   placement machines 145
manual assembly
   components 138-9, 144
manual soldering 181-2
manufacturing processes
   multilayer capacitors 48-50
   practical example 236-42
   resistors 40-42
   sample manufacture 234-6
Market Requirement 228-9
masking
   solder 32
materials
   cost comparisons 212
mechanical cycling tests 214, 215
MELF devices 34
   capacitors 54-5
   diodes 88, 89
   resistors 45-6
melting points
   solder alloys 165
meniscus
   solder inspection 189, 190, 191
metal-cored substrates 32-3, 253
metallisation (*see also*
         conductors; electroplating)
   ceramic substrates 24-7
metallised film capacitors 55-6
mica chip capacitors 54
Micropack 104
Microwave interconnection 248
miniaturisation 199-203
MiniMELF package 88
mixprint circuits 68, 81, 114-15, 225,
         230, 231
multilayer capacitors 46-54
multilayer PCBs 30, 247-9
Multiwire interconnection 248

naked chips 79-81
Nitto Kogyo Co.
   placement machines 154, 161
NTC thermistors 67

off-line programming 143
oil

wave soldering 178
on-line programming 143
optical automatic inspection
   placement 188
   substrates 186
opto-electronic devices 109-11
orientation
   of components 180-81
      see also design
organic substrates 27-32
oscillators 63-5
ovens
   solder drying 238-9

packaging
   integrated circuits 108
   passive devices 37-8
   semiconductor devices 91-2
   trends 13
packaging density 246-54
pad arrays 102-4
pads
   design rules 121-5
   soldering 120-21
   testing 193, 194
Panasert
   placement machines 150, 154, 155,
         159-60
Parylene
   protective coatings 197
passive devices 34-77
   history 4-7
pastes, solder *see* solder pastes
PCBs *see* printed circuit boards
performance characteristics (*see also*
         reliability)
   integrated circuits 105
   resistors 42-4
   semiconductor devices 90
phenolic/paper boards 28
Philips
   placement machines 157-9, 160-61, 240
photo devices 109-11
pick-and-place machines 144-61
   heads 142
pin grid arrays 102, 103
pin transfer
   adhesives 141, 174
   solder pastes 165
placement
   components 138-62, 167, 238, 239-40
      inspection 187-9
   machinery
      design restrictions 116-18
plastic film capacitors 7, 55-6
Plastic Leaded Chip Carriers (PLCC) 99,
         101, 137
plated-through holes
   and packing density 247-8
   ceramic substrates 23-4
   metal-cored substrates 33
   organic substrates 30, 247

plating (*see also* metallisation)
  additive process for PCB 28
  edge connectors 31
PLCC (Plastic Leaded Chip Carriers)
    99, 101, 137
polymers
  thick film circuits 32
polyimide substrates 19, 31, 208
porcelain enamelled steel
  heat dissipation 253
post cleaning 182-3
pot cores 11,12
power dissipation
  integrated circuits 105
  resistors 46
potentiometers 65-6
powered functional cycling 214
predrying
  solder pastes 167
preforms
  solder 172-3
pressure dispensing
  adhesives 141
  solder pastes 166
pretinning of substrate 31, 172
print and etch methods 24, 29, 247
printed circuit boards (*see also* substrates)
  cleaning 182-3
  design 113-37, 231-6
  history 3-4
  properties of materials 28
printing, screen *see* screen printing
programming
  placement machines 142-3, 144, 157-9
  protective coatings 195-8
  test machines 192
prototype manufacture 234-5
PTC thermistors 67-8
punching
  holes 18, 30, 247

quad flatpacks 100-102, 108
quality control 242
  capacitors 54
  integrated circuits 108-9
  resistors 42-4
  semiconductor devices 92-3

radial components
  PCBs 4
radiation
  infra-red soldering 168-9
reels
  tape-on-reel packaging 38, 91-2
reflow soldering 163-73, 238-9
  design restrictions 118-19, 120-21
  'footprints' 130-37
reliability 214-25, 243-4
  capacitors 54
  integrated circuits 108-9
  resistors 42-4
  semiconductor devices 92-4

repair work 194-5, 244
resin fluxes 183
resistance
  conductors 29
resistance heating
  soldering 171-2
resistors
  component/substrate
    compatibility 16
  history 5, 12-13
  passive devices 39-46
  soldering 'footprints' 130
  thick film substrates 26-7
resists *see* solder resists

safety precautions
  beryllia 22
  component placement 143
sample manufacture
  design evaluation 234-6
saturated vapour soldering 169-71,
    239, 240
Schlup, G. E., & Co.
  placement machines 145-6
screen printing
  adhesives 141
  solder pastes 166-7, 236-8
semiconductor devices
  active devices 82-94
sequential pick-and-place machines 144,
    147-57
Siemens Ltd.
  placement machines 154, 156
SIL packages
  3-dimensional layouts 251
silicon carbide
  substrates 22-3
silicon integrated circuits 8
silver
  alloys
    solder pastes 165
single jet wave soldering 177-8
sizes *see* dimensions
Small Outline Integrated Circuits 94-8
snapstrates 23
SO packages 10-11, 79
sockets 73, 74, 106-7
SOD packages 87-9
  soldering 'footprints' 135
software
  computer-aided design 127, 128
  placement machines 143, 144, 157-9
SOIC packages
  joint cracking 221, 223
  soldering 'footprints' 136
solder balls 167, 190
solder pads 120-21
  design rules 121-5
solder pastes 165-7, 236-8
  inspection 186-7
  placement machine functions 140-41
solder preforms 172-3

solder resists
  PCB manufacture 32
solderability
  resistors 43
soldering (*see also* flow soldering;
      reflow soldering)
  assisted by good design 39
  design restrictions 118-21, 130-37
  hand soldering 181-2
  inspection 189-92
  integrated circuits 105-6
  repair work 195
  semiconductor devices 90-91
  testing 218-19, 220-25
    SOT packages 10, 83-7, 79
soldering 'footprints' 132-4
spacing
  components
    design rules 123-4
  pads
    design rules 123
spattering
  solder 167, 172
specification
  for new product 228-9
spray coating 196-7
step stress testing 214, 236
substrates 15-33, 113
  (*see also* printed circuit boards)
  inspection 186
  packing density 246-9, 253
  unusual shapes 208-9, 249-53
subtractive processes
  substrates 29
surface tension effects 172
switches 73-5

TAB (Tape Automated Bonding)
      devices 104, 249
tantalum
  electrolytic capacitors 7, 57-9
taping
  packaging 37-8, 91-2, 108, 141
TDK Corporation
  placement machines 157, 159
temperature (*see also headings
      beginning* thermal . . .)
  operating range
    dielectrics
      capacitors 52-3
temperature coefficients
  dielectrics
    capacitors 52
  resistors 43-4
test pads 193, 194
testing
  component placement 143
  design restrictions 121
  electrical 192-4
  for reliability 214-15, 216-25
  integrated circuits 108-9
  practical example 241-2

samples 235-6
semiconductor devices 92-3
thermal coefficient of expansion 215
thermal cycling
  integrated circuits 106
  tests 214, 215, 218-19, 224-5
thermal dissipation 253-4
thermal mismatch
  component/substrate compatibility
      16, 98-9. 215-16
thermionic valves 1-3
thermistors 67-8
thick film circuits
  ceramic substrates 10, 20, 25-7
  comparison with SMA 206-8
  history 9-10
  organic substrates 32
  resistor networks 46, 47
thickness
  conductors 29, 247
  substrates 18, 231
thin film circuits
  ceramic substrates 24-5
  comparison with SMA 206-8
  history 9,10
three-dimensional design 249-53
through connections *see* 'via' holes
tin-lead solders 165
TO packages 89
tolerances
  multilayer capacitors 47-8
  resistors 39
  temperature coefficients
    dielectrics
      capacitors 52
'tombstone' effect 172
transfer printing
  adhesives 141
transformers 62
transistors
  history 8
trimmer capacitors 66-7
trimming
  resistors
    thick film substrates 26-7
tweezers
  manual assembly 139

ultra-violet curing
  adhesives 173-4
undercutting
  etching 29, 247
unencapsulated chips 79-81
Universal Instruments Corp.
  placement machines 146-7, 150-51
upgrading
  of product 244-5
User Requirement 228-9

vacuum pipettes
  manual assembly 139

valves
electronics history 1-3
vapour phase soldering 169-71,
239, 240
variable capacitors 12
variable inductors 62
variable resistors 12
Very Small Outline packages 97, 98
'via' holes 30, 247, 135
vibration
component feeding methods 141-2
viscosity effects 165, 173
visual inspection
soldering 189, 191
voltage coefficient
resistors 44
voltage rating

ceramic capacitors 54
VSO packages 97, 98

washing
after soldering 183
wave soldering 118-19, 119-20, 130-36,
177-9, 181, 241
width
conductors 29, 247
pads 123

X-ray inspection
solder joints 191

zero defects 94, 109
zero-hour quality 93-4
zero-ohm jumper 40

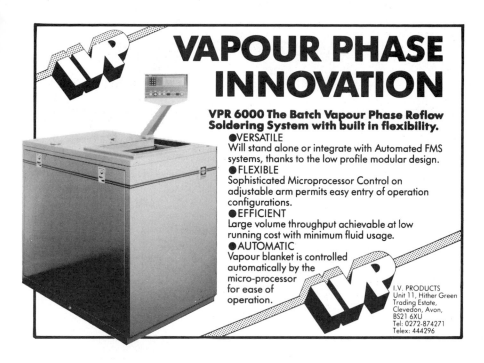

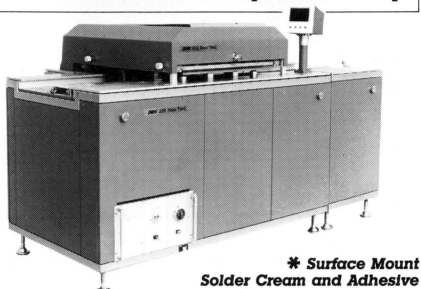

A Cookson Group Company

TAKE A CLOSER LOOK AT FRY'S

The name on the finest range of solders, fluxes and flowcreams for the electronics industry.

With many years of experience Fry's are always in the forefront of new soldering technology, supplying the quality products demanded by today's electronics industry.
For example:-
Fryflow and **Superspeed** – two top quality resin cored solder wires.
Super LDC alloy solder and **Chipflow Flux** for surface mounted components. Also **Fry's Flowcream** – the solder cream for screen printing and dispensing.

FRY'S METALS LTD
Tandem Works, Merton Abbey, London SW19 2PD.
Telephone 01-648 7020. Telex 265732.

525

CLEANING AND CONTAMINATION OF ELECTRONICS COMPONENTS AND ASSEMBLIES

by B. N. Ellis

Pages—365 + xxi;   Tables—17;   Figures—99;   References—159;   Size—23 × 15 cm.
ISBN 0 901150 20 7

This book is a practical guide for all persons who are in any way concerned with cleaning or contamination control of either components or assembled circuits. It is also useful to those specifying components and, in particular, printed circuits. It equally covers the theoretical side to a sufficient extent to allow the average engineer or technician who is not a specialist in the field to understand the mechanisms involved in contamination and cleaning.

As a reference book, the text is divided into seven parts, logically divided into some thirty chapters illustrated by photographs, line drawings, graphs and tables. Each chapter has its own reference list. The introductory section comprises an historical background to cleaning in the electronics industry, a very complete chapter of definitions of all the terms employed and in the particular context, a short chapter on units employed, a theoretical treatise on the mechanics, physics and chemistry of cleaning (in simple terms) and one on the cost of cleaning.

The second part comprises some seven chapters cataloguing the diverse ways that contamination can occur in the electronics industry, whereas the third part describes, over three chapters, what effects contamination can have during the various manufacturing processes of components and assemblies and over the whole of their subsequent lives. Part 4 will be considered as being the most important by some production engineers because its four chapters describe all the currently used methods of cleaning and flux removal for the small, medium or large user with considerable detail on the products usually employed.

The fifth section deals with ionic contamination control. The first two chapters discuss respectively the American military specifications and the new British DEF standards. The third one gives a general view of the different instruments commercially available for measuring or detecting ionic contamination. The last chapter of this section gives an insight into some aspects of the theory of ionic contamination measurement and the solutions used for it. The next part treats the detection and measurement of contamination by other methods, with particular emphasis on non-ionic contaminants. Insulation resistance measurement is discussed in a separate chapter of this section. The last part, divided into three chapters, relates to the particular problems imposed by the use of surface mounted components and solder creams and pastes.

It is felt that this book will become a valuable reference work for the bookshelves of all companies involved in any aspect of electronics, particularly component, printed or hybrid circuit manufacturers or assemblers.

May be ordered direct from

ELECTROCHEMICAL PUBLICATIONS LTD,
8 BARNS STREET, AYR KA7 1XA, SCOTLAND. TEL: (0292) 263281. FAX: (0292) 284719

SOLDERING IN ELECTRONICS

by R. J. Klein Wassink

Pages—470+xviii; Tables—57; Figures—301; References/Bibliography—686; Size—23×15 cm.
ISBN 0 901150 14 2

The book 'Soldering in Electronics' has been written to cover soldering as a technique for mass production of electronic connection, although many items treated will be useful for other branches of soldering technology.

Chapter 1 deals with the interaction of the various aspects which determine the result of the soldering operation, namely the quality of the soldered joints. It is explained that the desired goals can only be achieved if soldering is treated as a coherent system, all separate aspects of which are to be well prepared, often long before the actual moment of the soldering operation. This way of thinking has many implications for the other chapters, such as on solderable coatings, solderability assessment, and joint design.

The book provides a combination of background knowledge and information for practice. In doing so it helps to clarify many points of basic discussion among soldering experts, such as wetting and dewetting, thermal influences, solderability and methods of its assessment, metallurgy of alloys and coatings, and process conditions. On the other hand it contains a wealth of detailed and practical information on soldering alloys, fluxes, coatings, soldering methods, pattern design and criteria for joint inspection. A separate chapter is devoted to the thermal aspects of soldering, a subject that is seldom discussed in literature despite its utmost importance for soldering.

It is this combination of theory and practice which makes the book indispensable to a broad group of readers with a responsibility for electronics manufacturing, both in R&D and production, such as design engineers, process engineers, production engineers, quality engineers, metallurgists and chemists.

At all relevant places in the text, literature references are included for those who wish to further their knowledge on the items discussed. Comprising over 400 references (listed in alphabetical author order) on soldering, the book provides an excellent survey of the existing literature. Moreover, a bibliography is offered of all books on soldering ever published, as well as a list of soldering items in several languages.

May be ordered direct from
ELECTROCHEMICAL PUBLICATIONS LTD,
8 BARNS STREET, AYR KA7 1XA, SCOTLAND. TEL: (0292) 263281. FAX: (0292) 284719

Brazing & Soldering

Published in association with the British Association for Brazing and Soldering

CONTENTS

No **10**

**Spring
1986**

Editorial   2

Contributors   3

The Mechanical Properties of Soldered
Joints to Surface Mounted Devices
by S. P. Hawkins, C. J. Thwaites and M. E. Warwick   4

A Practical Comparison of Wave Soldering Equipment
for Surface Mounted Components
by C. J. Brierley and J. P. McCarthy   7

Diffusion Bonding Ceramics with Ductile Metal
Interlayers
by M .G. Nicholas   11

The Relationship of Cleaning Copper to Solder
Levelling—Characterising the Surface of Coppper
by T. Sulzberg, F. Shubert and W. D. Brewster   14

Some Properties of Soldered Joints Made with a
Tin/Silver Eutectic Alloy
by J. London and D. W. Ashall   17

Soldering of Surface-mounted Devices—New
Tasks and Their Solutions
by R. Strauss   21

Gaps in High Temperature Brazing
by G. R. Bell   24

Reliable Solder Paste Quality Control
by B. Roos-Kozel   35

Some Metallurgical Aspects of SMD Technology
by C. J. Thwaites   38

Using Differential Scanning Calorimetry to Determine
the Percent Tin in High-lead Solders
by J. G. Ameen and V. G. Veeraraghavan   43

BABS News   48

Special Reports   51

Industry News   55

New Products   57

Classified Buyers Guide   59

For information on subscriptions to *Brazing & Soldering*, please contact the publishers:

**WELA PUBLICATIONS LTD, 8 Barns Street, Ayr KA7 1XA, Scotland
Telephone: (0292) 283186   Fax: (0292) 284719**

Circuit World

Journal of the Institute of Circuit Technology

Volume 12 Number 3 1986

CONTENTS

Editorial	2
Contributors	3
Prepreg Testing—Quick, Informative, Reliable A Critical View by W. Schiffer	4
Advanced Solder Masks by M. R. Sculpher	9
The Offset Printing Method of Inner Layers—One of the Improved Methods for PCB Manufacturing Technology by A. Toushinsky and V. Evjenko	12
Predicting Reliability of LSI Printed Circuit Carriers by J. J. Tomaine	14
The Reliability of Surface Mounted Solder Joints Under PWB Cyclic Mechanical Stresses by C. J. Brierley and J. P. McCarthy	16
Use of Thermoplastics and Injection Moulded PWBs by D. Frisch and E. Cleveland	20
Wave Soldering of Surface Mount Components by W. K. Boey and R. J. Walker	25
Surface Mount Technology for High Reliability Telecoms Applications by K. Taylor	27
Flexible Circuit Materials for Use as Interconnections in High Speed Digital Systems by S. Gazit	30
Screenprinting in the Printed Circuit Board Industry by C. I. Wall	36
Literature Abstracts	41
Book Review	42
Company Profile—G. Bopp & Co. AG, Wolfhalden and Zürich, Switzerland	43
Company Profile—New England Laminates (UK) Ltd, Skelmersdale, Lancashire: Nelco SA, Mirebeau sur Bèze, near Dijon	63
Company Profile—Melchert Elektronik, Cologne (Lövenich), W. Germany	67
Internepcon Production Show and Conference	71
Circuit Technology '86 Joins 'The Week'	75
ICT News	80
International Institute News	81
Industry News	85
New Products	89
Classified Buyers Guide	102

For information on subscriptions to *Circuit World*, please contact the publishers:

WELA PUBLICATIONS LTD, 8 Barns Street, Ayr KA7 1XA, Scotland
Telephone: (0292) 283186   Fax: (0292) 284719